GEORGE F. SOWERS

MW00709768

# BUILDING ON SINKHOLES

Design and
Construction of
Foundations in
Karst Terrain

Published by

**ASCE**
*PRESS*

American Society of Civil Engineers
345 East 47th Street
New York, New York 10017-2398

**ABSTRACT:**

Based on over 50 years experience in geotechnical engineering, this book summarizes the mechanisms of sinkhole formation in limestone (or karst) terrain. The author also provides methods for overcoming sinkhole-related failures and for avoiding or minimizing future sinkhole collapses that impact on human activity. Professor Sowers also discusses site investigation, as well as the design and construction methods that are appropriate for building foundations in areas where sinkholes are likely to develop. It is written for engineers and geologists, in addition to other professionals who work together to solve sinkhole problems.

Library of Congress Cataloging-in-Publication Data

Sowers, George F.
Building on sinkholes : design and construction of foundations in Karst terrain / George F. Sowers.
p.        cm.
Includes bibliographical references.
ISBN 0-7844-0176-4
1. Foundations—Design and construction. 2. Sinkholes. 3. Limestone. 4. Karst. I. Title.
TA775.S645 1996      96-17831
624.1'5—dc20          CIP

The material presented in this publication has been prepared in accordance with generally recognized engineering principles and practices, and is for general information only. This information should not be used without first securing competent advice with respect to its suitability for any general or specific application.

The contents of this publication are not intended to be and should not be construed to be a standard of the American Society of Civil Engineers (ASCE) and are not intended for use as a reference in purchase specifications, contracts, regulations, statutes, or any other legal document.

No reference made in this publication to any specific method, product, process or service constitutes or implies an endorsement, recommendation, or warranty thereof by ASCE.

ASCE makes no representation or warranty of any kind, whether express or implied, concerning the accuracy, completeness, suitability, or utility of any information, apparatus, product, or process discussed in this publication, and assumes no liability therefore.

Anyone utilizing this information assumes all liability arising from such use, including but not limited to infringement of any patent or patents.

Photocopies. Authorization to photocopy material for internal or personal use under circumstances not falling within the fair use provisions of the Copyright Act is granted by ASCE to libraries and other users registered with the Copyright Clearance Center (CCC) Transactional Reporting Service, provided that the base fee of $4.00 per article plus $.25 per page is paid directly to CCC, 222 Rosewood Drive, Danvers, MA 01923. The identification for ASCE Books is 0-7844-0176-4/96/$4.00 + $.25 per page. Requests for special permission or bulk copying should be addressed to Permissions & Copyright Dept., ASCE.

Copyright © 1996 by the American Society of Civil Engineers,
All Rights Reserved.
Library of Congress Catalog Card No: 96-17831
ISBN 0-7844-0176-4
Manufactured in the United States of America.

# TABLE OF CONTENTS

# PREFACE

This book reflects a half century of geotechnical engineering. A significant part of that time was involved with investigating damage caused by sinkholes to buildings, roads, dams, and other structures. This evolved into developing the most effective ways to prevent sinkhole problems during planning, designing, and constructing new projects in areas prone to sinkhole development.

The book summarizes my observations and research into the mechanisms of sinkhole formation and the natural and human forces that drive the mechanisms. Fifty years ago there were many misconceptions of the nature of sinkholes. This has changed. There are new and better ways to explore site conditions. Extensive and deep construction excavations have made it possible to examine the rock-soil interface directly. New data have helped to clarify the relation between groundwater changes, rock solution, and soil erosion. This has made it possible to correct most failures that have occurred and to avoid or minimize future sinkhole collapses that impact on human activity. The text reflects this progress. It describes, in some detail, for the non-geologists as well the non-engineers, the mechanisms of sinkhole development and their impact on humans. However, geologists can learn from the engineering observations based on construction experience and the engineers can learn form the geologists and hydrologists. The text includes site exploration, familiar to both engineers and geologists, but its emphasis on the particular techniques that are most useful in areas underlain by solutioned rock should be helpful to both. It also points out the similarities of sinkholes over solutioned limestone and those over openings in other rock formations, such as lava and even over sewers and mines. A third part of the text deals with design and construction measures appropriate for building structures in areas where sinkhole activity is liable to develop. It also describes where some measures that are sometimes employed not likely to succeed. The text does not include water-retaining structures. Although the present work reflects the author's experience with dam design, construction, and remediation work, material on that specialty would nearly double the size of the present work; therefore the preparation and design for dam foundations are not discussed.

Many people have contributed to this work by their insights when working with the author in solving foundation problems and failures. Others have reviewed drafts of this text and made many helpful suggestions. Most are or have been co-workers with Law Engineering and Law Environmental Services, particularly, Randy Knott, Clay Sams, Luther Boudra, and David Wheeless, who reviewed earlier versions of the text; their comments and suggestions improved it greatly. Dr. Janet Sowers Horn, the author's daughter, a consulting geologist and an expert in cave geology, made many helpful suggestions as well as provided insight into the geometry of the underside of sinkholes. Frances Sowers, a retired hydrologist with TVA and The U.S. Corps of Engineers, the author's wife, has shared the field examination of sinkholes and sinkhole damage with me for the past 53 years. She contributed her critique of the written text, improving the logic and the flow of the wording, as she has with my previous works. Finally, Robert Alexander, draftsman with Law Environmental Services turned sketches into finished drawings, and secretaries Betsy Reed, Peggy Farley, and Chris Shattuck helped smooth rough drafts into the completed manuscript. The author thanks all for their help in making this book a reality.

The two technical reviewers, Dr. Dan Brown Professor of the Department of Civil Engineering, Auburn University, and Dr. Allen Hathaway, Professor of Geological Engineering, The University of Missouri at Rolla, made significant contributions to the scope of the text and the flow of the words. The subject is so broad and at the same time technically so complex that some compromises were necessary to challenge the diverse persons that are involved with sink hole problems. It is my hope that the book will be a catalyst to bring this diverse expertise together to solve the problems instead of debating endlessly who is responsible and who should pay (although those problems also require answers).

# CHAPTER 1

# INTRODUCTION

## 1.1 HOUSE IN A SINKHOLE

After 2 years of unusually dry weather, a family was enjoying the sound of the latest of several intense thunderstorms at supper time. Their one-story house had been built in a new subdivision approximately 5 years previously on former farmland, just north of the town of Bartow, Florida. Bartow is in the central part of the state, approximately 35 mi (60 km) east of Tampa, in a part of a broad low ridge that forms the backbone of the Florida peninsula.

Suddenly, the house shuddered. The family members ran out the front door to find a 20-ft (6-m) diameter hole near one corner of the house. The hole enlarged during the night, but the house, although bent, remained intact. During the next day, the city dumped several truckloads of sand from a nearby pit into the hole, filling it to the original ground surface and temporarily stopping the subsidence. Subsidence continued erratically for several days, followed by refilling in an attempt to maintain the ground surface level. This only retarded the house settlement.

After several days of subsidence and filling, a second sudden dropout occurred accompanied by the loss of one end of the house in an open hole (Fig. 1.1). By this time the family's possessions had been removed and the family was housed elsewhere.

The uppermost soil in the subdivision consists of poorly consolidated silty sands and slightly clayey silty sands, 30 to 40 ft (9 to 12 m) thick, underlain by an irregular layer of highly plastic clay and clayey sand containing calcium phosphate nodules from 1 to 10 mm in diameter. Still deeper is a porous limestone with irregularly spaced, steep solution-enlarged fissures. During the dry years proceeding the failure, the local groundwater table had been further depressed by wells supplying domestic water to the city. In addition, large volumes of water had been utilized for processing the phosphate pebbles obtained by strip mining of the phosphate-bearing clay

FIG. 1.1.    Sinkhole Beneath a Corner of a House in Bartow, Florida Related to Water Table Depression from Water Well Pumping (Photograph by the Polk County Democrat with Permission of the Publisher, S. L. Frisbe IV)

beneath the sandy overburden. The combination of a depressed groundwater level and rainfall infiltration at the ground surface caused erosion of clayey sand inclusions in the phosphatic clay layer and of the silty sand above. Eventually, a dome-shaped cavity developed in the silty and clayey silty sand beneath the house. When the soil cavity collapsed, the corner of the house sank. Unfortunately, several other homes in the same subdivision suffered similar fates over a period of 3 weeks.

## 1.2 DECANTING BASIN DROPOUT

Approximately 600 mi (1000 km) northwest of the house-eating sinkhole, nearly three decades later and in an area underlain by an entirely different limestone formation, circular dropouts appeared suddenly in the bottom of an elongated sedimentation-decanting basin for fly ash at a large industrial plant (Fig. 1.2a). The basin had been constructed by excavating 25 ft (8 m) deep in stiff residual clay that was approximately 40 ft thick over level bedded limestone. The residual soil in the basin bottom and on the 2H to 1V side slopes had been compacted by rolling with a sheepsfoot roller.

(a)

(b)

*FIG. 1.2. Sinkholes in an Ash Decanting Basin at an Industrial Plant in Northwest AL: a. Aerial View of Sinkholes in the Left Basin, at the Arrow; b. Ash Basin Bottom with the Largest Sinkhole and a Protective Workers Cage Supported by a Crane Hoist that Protects the Inspector in Case of Further Collapse*

It was then blanketed with two 8-in. (200-mm) layers of clay compacted to 95% of the Standard Proctor maximum. For nearly 25 years, the basin had functioned with no significant leakage or maintenance. The settled solids were removed periodically as needed, using a clamshell bucket and placed in a solid waste disposal fill 0.5 mi (0.8 km) distant.

The largest dropouts were near the basin center, one approximately 4 ft (1.2 m) in diameter and the second approximately 6 ft (1.8 m) in diameter. Both were found to be partially filled with ash. Excavation of the dropout areas disclosed a possible connection between the two, several feet below the basin bottom and above a deep vertical hole with some limestone outcropping in the bottom. Fortunately, the designer had the foresight to require two such basins, to allow for the cleaning of one while the other was filling. The dropout holes were discovered during cleaning one basin; the second basin permitted the plant to operate while repairs were being made to the first.

Both failures are typical of the problems that occur in areas that are underlain by dissolutioned limestone. Although the two examples occurred in widely separated locations at different times, they have two things in common. First, the underlying limestone rock was blanketed with a cover of soil that obscured any fissures or cavities in the rock. Second, the natural groundwater and surface infiltration regimen had been changed significantly. The examples are by no means unusual.

### 1.3 AN EXTRAORDINARY SINKHOLE

In May, 1981, a dropout or *sinkhole* 300 ft (100 m) in diameter and 50 ft (15 m) deep to the water surface developed during a 2-day period in downtown Winter Park, FL (Fig. 1.3). It destroyed two city streets, a municipal swimming pool, a house, an automobile repair shop, and several underground utility lines. Several automobiles parked behind the repair shop slid down the unstable slope and came to rest just short of the water. They were later recovered reasonably intact.

Similar but smaller dropouts occur as frequently as hundreds (or more) yearly in some parts of the United States, causing significant property losses. Fortunately, most of the subsidences or dropouts occur so slowly that there is rarely loss of life. However, there is potential for casualties when structures are built in areas subject to such subsidences and dropouts, unless measures are taken to minimize the groundwater depletion and the increases in surface water infiltration that precipitate them.

### 1.4 EXTENT OF SUBSIDENCES AND DROPOUTS RELATED TO ROCK SOLUTION

By far the greatest number of such problems worldwide occur in regions underlain by soluble rock formations, principally limestones and dolomites.

(a)

(b)

FIG. 1.3. A Sinkhole in Winter Park, FL, May, 1981, 300 ft (91 m) in Diameter and 60 ft (18 m) Deep to Water: a. Sinkhole Throat, Damaged Auto Shop on the Rim and Vehicles from a Parking Area Scattered on the Slope of the Hole. Viewed from the Sinkhole Rim Looking Toward the Top Right Corner of Fig.1.3b; b. Aerial View with the Auto Shop on Rim at Top of the Picture, Vehicles on the Slope Below, Two Broken Intersecting Streets Near the Photo Center and a Broken Swimming Pool at the Bottom Center

Percolation of water through fissures in the otherwise sound rock enlarges them, producing cavities of varying sizes, shapes, and extent in the rock. This is followed by ravelling and erosion of soil overburden into the rock cavities. New and different cavities with dome-shaped tops are generated in the soil overburden. If the process continues, the soil cavity roof first subsides and eventually collapses. The collapse of the roof of a shallow cave in the limestone can also cause a subsidence or dropout at the ground surface. However, subsidences and dropouts from the collapse of caverns in the rock are infrequent during the lifetime of humans, although not in geologic time spans.

Similar subsidences and dropouts can develop over any soil or rock cavity regardless of the nature of the rock formation. These cavities include lava tubes, man-made tunnels, mines, and sewers. This book focuses on subsidences of areas underlain by limestones and dolomites; however, it briefly describes similar dropouts from natural and man-made cavities below the ground surface.

### 1.5  DEFINITION OF LIMESTONE AND SUBSIDENCE-DROPOUT TOPOGRAPHY

Unfortunately, the terminology for cavity-induced subsidence is somewhat confused; similar materials and phenomena are known by different names by the different professions involved (i.e., geologists, hydrologists, geomorphologists, engineers, and miners), in different languages and even in different regions or nations that use the same language, such as the United Kingdom and the United States.

The author employs the terms commonly used in engineering and construction in the United States by non-specialists. Alternate terms used by specialists are defined in Section 3.11 to aid the reader in understanding writings aimed at specialists, particularly those in other disciplines such as geohydrology and geology.

The principal rock formations involved in dropout and collapse activity, *limestones* and *dolomites*, are largely composed of the minerals calcite (calcium carbonate) and dolomite (calcium-magnesium carbonate). Some specialists prefer *dolostone* as the rock name to the traditional dolomite because the suffix "-ite" is more properly applied to mineral rather than rock. However, the term dolomite is more widely used in the United States than dolostone. The engineering characteristics of limestones and dolomites as related to construction problems and to catastrophic collapse are so similar that the term "limestone," as used in this text, will include both, unless specific mention is made of dolomite. Some writers use the term *carbonate rocks* or carbonates to encompass both limestone and dolomite. Metamorphosed limestone, *marble*, is included in this extended limestone definition because the engineering properties that are involved in the subsidence and dropout problems are similar to those of the unmetamorphosed rock. (Where specific formations are involved, such as the *Knox Dolomite* of Alabama, East Tennessee, and Virginia, the accepted name for the formation will be used, although that formation may include other rocks.)

The term *karst* is widely used to characterize the topography produced by subsidences and dropouts in the ground surface that develop above limestones from various mechanisms involving water. It comes from the Karst area of what was northwestern Yugoslavia; however, the term often has been applied to dropout topography associated with limestones that only superficially resembles that of the prototype region. *Limestone terrain* includes the total range of subsidence and dropout topography, but does not imply any particular degree of development. This book will restrict the term *karst* to subsidence and dropout topography that displays most of the more prominent features of solution-induced topography as described by Fairbridge (1968, p. 582): numerous closed ground surface depressions or dropouts and the absence of smaller surface water drainage. Specific solution features will be defined in Section 3.11.

# CHAPTER 2

# FORMATION OF LIMESTONE DEPOSITS

## 2.1 DEPOSITION

Most limestones form in water, largely in the sea, but sometimes locally in fresh or brackish water. Calcium and magnesium carbonates form from the chemical weathering of rock minerals containing calcium and magnesium. Magnesium and calcium carbonate are also present as rock minerals. These become soluble bicarbonates in the presence of water containing dissolved carbon dioxide (carbonic acid). Slightly soluble sulfates are formed by the attack of dilute sulfuric acid produced by organic decay or weathering of some sulfide minerals. These new soluble minerals are leached from their point of formation and transported in solution by groundwater and surface runoff until some change in the chemical and biologic environment in the water causes precipitation.

The environment for bicarbonate precipitation out of water is particularly favorable in warm shallow seas, such as the Bahamas Banks east and south of Florida, where the process can be readily observed today (Fig. 2.1). There are two mechanisms that proceed simultaneously: 1) direct chemical precipitation and 2) biologic absorption by marine organisms followed by sedimentation of the organisms' skeletons. Because the solubility of the carbon dioxide in water decreases with increasing temperature, warm water favors direct precipitation in the form of silt-sized particles, some of which grow by accretion in layers about a small nucleus to form spheres (Fig. 2.2a). Similarly, warm water promotes the growth of the carbonate secreting organisms, such as shell fish (Fig. 2.2b) and corals (Fig. 2.2c), whose skeletal remains add to the calcium carbonate fragments on the sea bottom and which may be swept onto nearby beaches (Fig. 2.1b). The predominant precipitation mechanism appears to be biologic. Some algae utilize calcium and magnesium carbonates in their growth (Fig. 2.2d). The remains of these organisms become lime muds, sands, and gravels, such as those that form

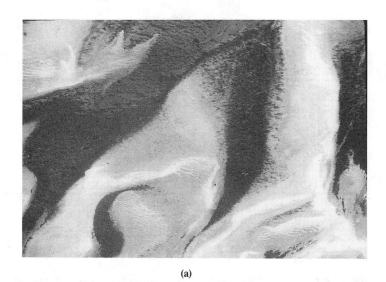

(a)

(b)

*FIG. 2.1. Biologic and Chemical Precipitates of Calcium Carbonate in the Bahama Islands: a. Newly Formed Particles on the Sea Bottom, Swept into Submerged Sand Bars by Submarine Currents in Approximately 30 ft (9 m) of Water; b. Sand Composed of Shell Fragments, Coral Fragments and Carbonate Precipitates From a Beach Adjoining the Sand Bars of Fig. 2.1a*

(a)

(b)

FIG. 2.2. *Photomicrographs of Young (Less Than 5 Million Years) Incompletely Consolidated Limestones: a. Spherical Precipitates, oolites, 0.2–0.3 mm in Diameter Forming an Colitic Limestone Near Miami, FL; b.* Coquina *(shell) limestone with particles 1-5 mm wide, north of Miami, Florida. c. Coral Limestone (1-mm Scale Marks), South of Miami, FL; d. Algae-precipitated Limestone, California (Photo by Janet Sowers Horn)*

(c)

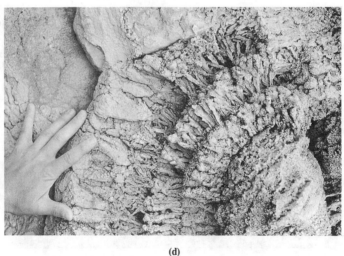

(d)

*FIG. 2.2.   (Continued)*

the submarine and surface sand bars of the Caribbean region. Bone accumulations are largely insoluble calcium compounds. Intact and broken remains
carpet the shallow sea bottom and accumulate in coral reefs and sand bars.

   Fresh and brackish water deposits of carbonates are similar to those of
the sea, although the environments are somewhat different. The deposits are

usually less extensive near the shore in non-ocean materials. The water turbulence is usually less, which reduces the mixing of different types of sediments; therefore, the range in different depositional features is usually narrower.

Hydrothermal action also produces limestones. Hot springs and geysers precipitate locally thick accumulations of calcium and magnesium carbonates and, to a smaller extent, sulfates. Such deposits often occur in areas of igneous and metamorphic rocks where limestones are not usually expected. Evaporation of lakes in arid regions also produces some deposits of limestone, usually interlayered with other evaporites, such as hydrous calcium sulfate (*gypsum*), anhydrous calcium sulfate (*anhydrite*), and sodium chloride (*halite* or salt).

There are variations in the composition of the carbonate sediments as the sedimentation environment changes. Limestones are often interlayered with *clastic* sediments (mineral particles from weathering of rock). These sediments range from clay and silt (the more common) to sand and gravel (less common). Sometimes the carbonate sediments are intermixed with clastic sediments forming *arenaceous* (sandy) limestones and *argillaceous* (clayey) limestones to calcareous (or dolomitic) sandstones and calcareous (or dolomitic) claystones. In deeper water, the materials are deposited in horizontal layers, or strata, with each layer of similar grain size and mineralogy. Near shore, where wave action and wave currents dominate, the sediments become irregularly mixed, then resorted. They are finally deposited in parallel ridges as shore bars in dipping or cross-bedded layers. In warm climates, coral reefs, sometimes more than 100 ft (33 m) thick and as wide as 1 mi (1.6 km) also accumulate near shore. The larger voids in the coral framework of calcium carbonate often fill with finer limestone and clastic sediments washed into them by wave and current action. Structural deformation, from local landsliding to tectonic upheaval, add other dimensions to the variations, producing warping and folding, accompanied by cracking from tension and shear.

## 2.2 INDURATION AND ALTERATION

Following deposition of the carbonate sediments, several processes proceed more or less simultaneously: 1) *consolidation* (densification or compaction) from the weight of the accumulating sediments, 2) re-solution and action exchange of the carbonate minerals, and 3) reprecipitation. The consolidation, which includes particle distortion, fracture, and rearrangement, reduces the porosity (void ratio) of the sediment. It also may produce some pressure welding between the particles at their points of contact. The re-solution, termed *dissolution*, occurs from water circulating through the sediment pores accompanied by changes in the water chemistry and temperature. Solid calcium carbonate temporarily becomes soluble calcium bicarbonate. The bicarbonate may be reprecipitated nearby in the same sedimentary mass with continuing changes in the solution-precipitation environment, bonding the particles together or carried away to be deposited elsewhere. There may be further changes. Magnesium carbonate is less soluble than calcium carbonate; therefore, if magnesium bicarbonate is present in the percolating water, it may replace the calcium in a process known as *dolomitization*. The calcium becomes soluble bicarbonate and is freed to be deposited elsewhere.

Silica (silicon dioxide) is also soluble in water, although to a much smaller degree than the carbonates. However, the solubility increases with an increase in pH. The less soluble silica can replace the calcium and magnesium carbonates, sometimes retaining the texture of the carbonates in the form of microcrystalline forms of silica known as *flint* and *chert*. The dissolved silica is also utilized by small organisms, such as *radiolaria*, for their "skeletons." Accumulations of this biologically precipitated silica also become consolidated into *chert* and *flint* within the limestone during the rock-forming processes. Some limestone formations contain extensive chert lenses. These are significant in construction because the silica is far harder than the limestones, resisting drilling, wearing construction equipment, and making it difficult to crush the rock in aggregate production.

The replacement mechanisms involved are complex, and the concepts of dolomitization and silica replacement in water after sedimentation are not agreed on by all experts. However, the magnesium content of newly formed carbonate deposits appears to be less than that of older deposits. Silica replacement appears to develop through water circulation in older carbonate rock pores, although it also occurs by sedimentation of organisms with silica skeletons. The precipitation bonding and the other chemical changes vary greatly from point to point and level to level within the rock mass.

The end result of consolidation, solution, and reprecipitation is the loss of the identity of most of the original particles; however, some fossils remain intact. Development of bonding between the particles is accompanied by a significant reduction in the porosity and void ratio of the mass, enabling the loose sediment to become a coherent rock.

## 2.3 ENGINEERING CHARACTER OF LIMESTONES

The engineering properties of limestone are probably the most varied of all sedimentary rock because of the many physical and chemical changes that accompany both their sedimentation and their post-sedimentary history. The particles that initially settle out have a wide variety of sizes, shapes, and micro-structures. Some are solid, some are hollow, some are near spherical, many are nearly equidimensional, and a few are sheet-like, elongated, or rod-like. The particles are sorted or heterogeneously mixed depending on whether they are transported and deposited in deep, relatively still water or in shallow, more turbulent water stirred by wave action near shore.

The spaces or voids between the intact rock particles and crystals are termed *primary porosity*. This is expressed quantitatively by either of two terms. Geologists prefer *porosity*, *n*, the percentage of the volume of *voids* or *pores*, $V_p$, compared with the gross volume of the mass including the pores, *V*. Geotechnical engineers prefer *void ratio*, *e*, the ratio of the volume of the *voids*, $V_v$ to the volume of the solid particles, $V_s$ (void ratio is more convenient in engineering calculations)

$$n = V_p / V \times 100\% \qquad (2.1)$$

$$e = V_v / V_s \qquad (2.2)$$

The primary porosities of some intact limestones recently deposited and unweathered are as high as 50% (a void ratio 1.0) and so low as to be not measurable by the routine laboratory tests. The primary porosity varies from point to point within the rock, reflecting varying degrees of homogeneity. Porosity also can vary with the direction through the rock. This difference in physical properties with direction is termed *anisotropy*. Anisotropy depends on the orientation of the elongated and sheet-like solid particles, the stress field accompanying consolidation, and variations in the depositional environment and the resulting stratification of the sediment.

The intact rock, including its voids, often contains fissures that interrupt the continuity of the mass. These are known as *secondary porosity*. Secondary porosity often occurs in more or less regular patterns, or *sets*, of nearly plane, parallel openings. Several sets in different orientations can occur at the same location, breaking the mass into blocks. Some fissures also occur randomly, as well as in broad zones of closely spaced crisscross (xxxx) fractures, such as those that accompany rock shear and faulting. *Vugs* are local large, often elongated pores, sometimes caused by arching between particles and fragments of widely different sizes.

Limestone strengths in unconfined compression can range between 200 lb/sq in. (1.5 MPa), the lower limit for *rock* by its engineering definition (Sowers, 1979, p. 68), to more than 20,000 lb/sq in. (150 MPa), greater than concrete.

## 2.4 SOLUBILITY

Limestones, gypsum and anhydrite, and salt (halite) formations are unique because they are very soluble compared to the other abundant rocks of the earth's crust. Although the gypsum-anhydrite and halite are by far the most soluble, their more-limited exposure near the ground surface (except for very arid regions) makes the solution of limestones of far greater general concern to engineering and construction.

In distilled water, limestone is somewhat less soluble than silica (usually considered to be insoluble). However, in water in which carbon dioxide is dissolved, or water with a pH lower than 7.0 (acid water), limestones become soluble. Carbon dioxide dissolved in water partially disassociates and forms carbonic acid, $H_2CO_3$. It is a weak acid but it reacts with calcite to form soluble calcium bicarbonate, $Ca(HCO_3)_2$, and with dolomite to form soluble magnesium bicarbonate, $Mg(HCO_3)_2$

$$H_2O + CO_2 = = H_2CO_3 \text{ (carbonic acid)} \qquad (2.3)$$

$$H_2CO_3 + CaCO_3 = = Ca(HCO_3)_2 \qquad (2.4)$$

A more detailed discussion of the chemical reactions involving water, carbon dioxide, and the carbonate minerals is given in such texts as *Geomorphology and Hydrology of Karst Terrains* (W. B. White, 1988).

The concentration of carbon dioxide dissolved in the water is a major factor in the rate of limestone dissolution. Although the solubility of carbon dioxide decreases slightly with increasing temperature, the effect is not dominant. Instead, the availability of carbon dioxide in the atmosphere in contact with the water and in the humus at the ground surface and the partial pressure exerted by the gas in contact with water are the most significant factors in limestone dissociation. Rainfall gains carbon dioxide falling through the atmosphere. Percolating into the topsoil, the water gains much more carbon dioxide produced by the respiration and decay of organic materials in the humus and topsoil. Other acids generated by organic decay and by decomposition of sulfide minerals increase the solution rate greatly. Acids from groundwater pollution also can increase solution locally.

The degree of saturation of the calcium and magnesium bicarbonate in water is also a factor in the solution rate: if the water is already saturated with carbonates, it cannot dissolve the limestone further. This is related to the circulation of the water in contact with the limestone. A long period of contact with the rock because of low water velocity or a long circuitous path increases the opportunity for saturation. Solution causes *denudation*: resembling "corrosion" of the rock surface. Because neither the seepage rate nor the rock composition is uniform, the result is an irregular, often grooved or fluted surface (Fig. 2.3).

At the present time, the theoretical knowledge of limestone dissolution does not make it possible to calculate the rate of rock surface removal or denudation from the chemistry of the rock and the water and the rate of flow

(a)

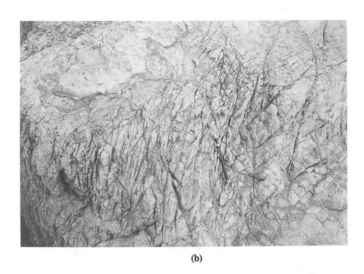

(b)

*FIG. 2.3. Solutioning of Limestone Surfaces (Denudation): a. Fluted Surface of a Solutioned Joint in a Well Indurated Limestone Exposed in a Highway Cut in East Tennessee (the Fluting Reflects Solubility Variations in the Horizontal Bedding); b. Solution Grooves,* rillkarren, *in a Limestone Exposure in North East Tennessee (Pencil Gives Scale)*

at a given locality. The mathematical models have not been verified by extensive field data and it is not possible to quantify the parameters that enter into those calculations. However, empirical studies do provide some guides.

The studies involve making measurements of the rock surface denudation or *deflation* at regular intervals on a variety of limestones in different environments. Published data compiled by J. N. Jennings (1983) and White (1988, p. 218) show rates varying from 5 to 200 mm per 1,000 years. The ranges of their data are shown in Fig. 2.4. The rates are greatest for a tropical environment, intermediate for temperate, and least for arctic environments. The average tropical rate is about twice that of the arctic. The denudation increases in proportion to the annual *runoff* (i.e., rainfall minus evapotranspiration) and with the cube root of the partial pressure of carbon dioxide in the air-water system. In addition, it increases with the decrease in water pH. For the climate of the eastern U.S. and western Europe, the rate is between 25 and 40 mm per 1,000 years. Thus, during a 100-year project life, the denudation in the eastern U.S. and western Europe is likely to be between 2.5 and 4 mm.

The mineral composition of the rock would be expected to be a factor. For example, the solubility of the mineral calcite in water containing carbonic acid is greater than that of dolomite. However, the solution of some

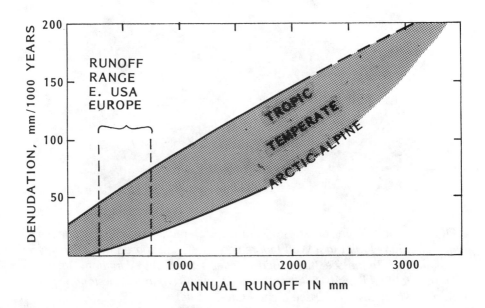

*FIG. 2.4.   Rates of Limestone Surface Removal by Solution (denudation) for Various Climates, based on Worldwide Data [Adapted From White (1988) by Permission]*

dolomite formations, such as the Knox Dolomite of the Southeastern U.S., is far more severe than that of some calcium carbonate limestones in the same environment. It appears that other associated minerals and local structural features, particularly fractures as well as differences in groundwater chemistry, are more important factors than the rock mineralogy in the solution of rock formations.

Significant limestone dissolution by sulfuric acid produced by hydrogen sulfide gas from organic decomposition or from natural gas percolation upward from far below has been identified in the Carlsbad Cavern area by Hill (1987). This may partially explain the extraordinary cavern dimensions. It is likely that similar acid solution is produced by polluted water and particularly by acid mine wastes. However, there is only limited documentation of the effect of these pollutants.

## 2.5 EFFECTS OF LIMESTONE SOLUTION ON PRIMARY POROSITY

There are three effects of limestone solution: 1) denudation or degradation of the rock surface, 2) enlargement of the primary (interparticle) porosity, and 3) enlargement of the secondary (fissure) porosity. This enlarged porosity, both primary and secondary, causes most of the serious structural and hydraulic problems that arise in limestone terrain.

A significant increase in primary porosity reverses the induration process. Visually, the fragments of the original sediments incorporated in the rock may reappear as discrete particles, with even larger voids between them than those that accompanied their deposition. The hydraulic conductivity of the rock mass thus increases greatly. This often aggravates the solution rate and thus further increases the porosity. For example, the primary porosity of the well-indurated, unweathered oolitic limestone underlying much of Miami, Florida, above the water table is approximately 15% (void ratio = 0.18). A few feet below the water table, where the circulation of groundwater is the greatest, the enlarged porosity is as high as 75% (a void ratio of 3). A 2-ft (600-mm) diameter core of limestone with this enlarged porosity is shown in Fig. 2.5. The porosity increase is 500% (a void ratio increase of nearly 1700%). The strength decrease from the loss of solids is somewhat proportional to the increase in porosity; the unconfined compressive strength of the rock in Fig. 2.5 is about 1/4 to 1/8 that of the relatively unsolutioned rock. *Vugs* (i.e., openings far larger than the rock particle size) that are interconnected with the water circulation system in the primary pores are also enlarged by solution. Sometimes large particles, such as fossils or chert, lie loose inside the vugs.

Under confined loading, the highly porous rock exhibits *compressibility*, similar to that of a loose sand. The rock mass volume decreases, accompanied by an equivalent reduction in porosity and void ratio. The

(a)

(b)

*FIG. 2.5.   Solution-enlarged Primary Porosity in Limestones From Miami, FL: a. Porosity 75% Approximately (popcorn rock); b. Porosity 30% Approximately (1-mm Scale Marks)*

author's limited settlement experience in South Florida suggests compress-ibility becomes significant in limestones, with well distributed primary po-rosity exceeding approximately 67% (a void ratio exceeding 2), caused by a combination of poor induration or the enlargement of primary porosity by solution. There are few data on limestone consolidation; however, it can be measured by consolidation testing of intact rock specimens. The weight of large structures or extensive fills caused settlements of 2 to 6 in. in situations such as described by the case history later in this section.

Most of the consolidation develops rapidly, and appears to stop a few weeks after the load is applied; however, there is some continuing settlement *(secondary compression)* at a steadily decreasing rate without a change in the loading, similar to that experienced in loose sands. The settlement mechanism appears to be largely the result of local crushing of the weakened bonds between the larger particles of the partially dissolutioned rock.

The strength and compressibility of the solutioned rock can be measured in laboratory tests similar to those for soils, but utilizing sufficiently large specimens to reflect the small-scale range in void variations and loss of bonding of the solutioned rock. In addition, a number of samples should be tested to reflect the large-scale rock porosity variations.

A college building under construction south of Miami, Florida, settled 2 to 4 in. (50 to 100 mm) during construction. The settlement developed simultaneously with placing crushed rock fill 2.5 to 3 ft (640 to 900 mm) thick over the building area. As the fill area enlarged, so did the area of settlement. Most of the settlement occurred rapidly; however, there was some continuing settlement, similar to secondary compression of soil, that continued at a decreasing rate. The additional settlement was less than 0.5 in. (13 mm) at a constant load during the following 6 months.

Field load tests confirmed that the limestone between 10 and 20 ft (3 to 6 m) below the ground surface had initial void ratios exceeding approxi-mately 2 (porosities exceeding 67%). The amount of consolidation under stresses similar to those produced by the fill weight was comparable to the measured fill settlement. Other cases of such settlement have occurred in the same area. The worst was approximately 6 in. (150 mm) of settlement of a multi-story office building supported on spread footings on limestone of low primary porosity approximately 15 ft (4.5 m) thick underlain by 10 to 20 ft (3 to 6 m) of limestone with high solution-enlarged primary porosity.

Solution that is focused in the secondary porosity, the bedding surfaces, and transverse joint fissures separates the rock mass into partially discontinu-ous blocks with irregular lenticular or sheet-like gaps or slots between them. Near the rock's upper surface, and at levels where groundwater circulation is the greatest, the rock slots are wider and the blocks are narrower. The solutioned surfaces are usually irregular, reflecting differences in the rock solubility as well as variations in water movement. Fig. 2.6 illustrates the

progressive solution enlargement of secondary porosity of joints and bedding surfaces. Figure 2.7 is a photograph of a highway cut in horizontally
stratified limestone with solution-enlarged joints and limited solution enlargement of some of the bedding surfaces.

As solution continues, the fissure gap widths increase and the block
thicknesses decrease, as illustrated in Fig. 2.6. Usually the solution is far
more rapid along the joint fissures than along the bedding surfaces, which
fit tightly against each other as long as the strata are horizontal and undistorted.
However, distortion of the rock formation accompanied by slippage along
the bedding sometimes creates large but somewhat discontinuous open
fissures. These enhance groundwater seepage along them so that the bedding surfaces also become solutioned, but seldom to the degree of the joint
fissure enlargement.

The gaps between the intact blocks are known by a number of names:
*slots, cutters, grikes,* and *bogas,* depending on their size or language of their
origin. In the interest of simplicity in descriptive terminology, this text will
refer to all solution-enlarged fissures as *slots;* the intact rock between will be
termed *blocks* if they are wider than high or *pinnacles* if they are higher than
wide.

Locally, solution can progress to the degree that continuous conduits or
passages, termed *caverns* or *caves,* develop. The size of these openings
appears to be limited only by the extent of the soluble rock, the ability of the
intact rock remaining to roof over the opening by either arching or flexural
strength, and the water circulation. The cavern systems of the Carlsbad, NM
(Fig. 2.8), and Mammoth Cave, Kentucky, areas appear to defy reason.
White (1988) and Ford and Williams (1989) discuss the formation of and the

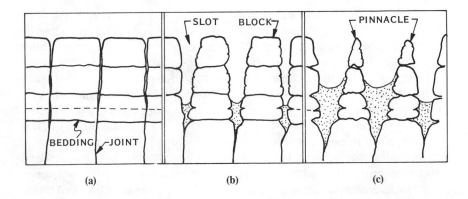

*FIG. 2.6. Solutioning of Vertical Cracks Developing Widening Slots and
Narrowing Blocks, and Eventually, Pinnacles as the Process Progresses: a.
Solution on Cracks; b. Slots Define Blocks; c. Blocks Narrow into Pinnacles*

FIG. 2.7. Solution-enlarged Vertical Cracks with Wide Blocks between Them and the Beginning of Solution of Horizontal Bedding Surfaces in a Highway Cut Near Nashville, TN (Red Clay Residual Soil Partially Fills the Slots)

terminology applied to caverns in technical detail. The reader should refer to these and similar treatises for further information on cave formation and geometry.

## 2.6 RESIDUAL SOILS FROM LIMESTONE

Limestones contain variable amounts of insoluble materials. When the soluble carbonates are dissolved and carried away in the groundwater, the insoluble components of the original rock are left behind. These insolubles include sedimentary gravels, sands, silts, and clays that were deposited simultaneously with the carbonates, chert, and flint (silt-sized to boulders) that formed in the deposited carbonate, and iron oxides precipitated from water and from weathering of iron minerals in the original sediments. These solids accumulate on the sides of the pinnacles and blocks; they partially fill the slots and cavities and blanket the rock surface with *residual soil*.

The proportion of insoluble materials varies greatly within any formation, depending on the environment of sedimentation and from one forma-

*FIG. 2.8.   Cottonwood Cave (Not Open to the Public) in the Carlsbad, NM, Vicinity (the Scale is Indicated by the Dark Shadows of Standing People) (Photo by Janet Sowers Horn)*

tion to another. Relatively pure limestones develop relatively clean slots with limited accumulations of residual soil; impure limestones usually weather more slowly than purer ones; however, the amount of insoluble materials for the same amount of original rock is greater. The thickness of residual soil that accumulates increases with the rate of rock solution; however, it is decreased by the erosion of the residual soil by surface-water runoff. Limestones with sufficient impurities to form large amounts of residual soil, but not enough to inhibit dissolution significantly, have the greatest rates of residual soil production. Their slots become partially filled and the denuding rock surface becomes blanketed with residual soil (Fig. 2.9). This residual blanket can be thicker than 100 ft (30 m), depending on the rate of rock weathering, the length of time it has been weathering, and the rate of erosion of the accumulating soil blanket. Of course, despite the residual soil accumulation, the elevation of the ground surface decreases during the process because each increment of soil thickness represents a much greater thickness of rock dissolved. The time required for the soil to develop can be estimated from the solution rates of Fig. 2.4 and the proportion of the insoluble materials in the rock. If the rock was 10% insoluble minerals, the rate of solid accumulation with free contact of the water would be at best only 10% of the rates shown on Fig. 2.4. Considering that the rate of flow across the rock with a soil cover is retarded by the soil, the accumulation rate would be a very small fraction of that for the free flow of water. Based on this quick simple model, it is obvious that the 100-ft residual soil thicknesses in the Eastern U.S. would have required more than 100,000 years to accumulate, without considering the lowered rate of solution caused by the accumulating soil cover that would protect the rock.

FIG. 2.9.   Dark Red Silty Clay Residual Soil over a Soft White Limestone in Haiti (There Is no Visual Evidence of Soil Stratification or Structure; the Transition From Rock to Residual Soil is Abrupt but Somewhat Irregular because of the Variability of the Primary Porosity in the Rock)

Most residual soils derived from limestone solution do not retain the relic structure of the original rock as do the soils derived from the weathering of other types of rocks. When the insoluble impurities are a small part of the rock volume (as is true of most limestones), the volume reduction accompanying solution does not occur uniformly because the insoluble materials are not uniformly distributed. Moreover, the solution proceeds three-dimensionally from the fissures inward. As a result the insoluble materials are rearranged and the stratification is virtually obliterated as the rock mass loses its identity. However, thick strata of chert, very sandy limestone, and similar concentrations of insolubles may remain as irregular bodies of impurity concentrations in the residual stratum. Calcareous mudstones and calcareous sandstones often retain some relic structure; however, the stratification often will be distorted, reflecting different proportions of insoluble impurities from point to point.

Solution weathering leaves the insoluble particles loosely packed. These newly formed residual soils are usually wet and soft. Their water contents sometimes exceed their liquid limits. When such residual materials form or accumulate in slots or cavities, they are likely to remain soft and paste-like unless the level of circulating groundwater drops below them. When exposed to air, they desiccate and stiffen, leaving the slot or cavity only partly filled and the rock surface of the slot or cavity plastered with stiff partially desiccated soil. The residual soils that accumulate above the rock surface are initially wet and pasty, but they consolidate under the weight of older residual soil above plus any deposited soils or even a structure above and become stiffer. Their stiffness is often augmented by iron oxide bonding.

Depending on the environment, the upper portion of the residual soil blanket undergoes further weathering from the ground surface down. This secondary weathering or *pedological weathering* develops a near-surface pedologic soil profile that reflects the near-surface environment. Often this includes localized partial desiccation, cementation, and hardening in the B Horizon, commencing 1 to 3 ft (0.3–1 m) below the surface in the eastern part of the U.S. and in regions of similar climates. Both dessication and cementing can increase the soil strength significantly in the uppermost 3 to 10 ft (1 to 3 m), as illustrated by the graph of N (the *standard penetration resistance*) as a function of depth in Fig. 2.10. [Standard penetration resistance is measured by blows of a 140-lb (63.7 kg) hammer dropping 30 in. (762 mm) required to drive a 1.4 in. inside diameter (ID), 2.0 in. outside diameter (OD) (35 mm ID, 51 mm OD) split tube sampler 1 ft (305 mm).]

The residual soils over limestones usually have the texture of sandy silty clays. They have a wide range of plasticity that reflects the clay content and clay mineralogy. If the parent limestone contained chert or flint, the soil will contain sand and gravel-sized fragments to boulders of chert. When chert is a significant portion of a particular limestone stratum, the chert may be

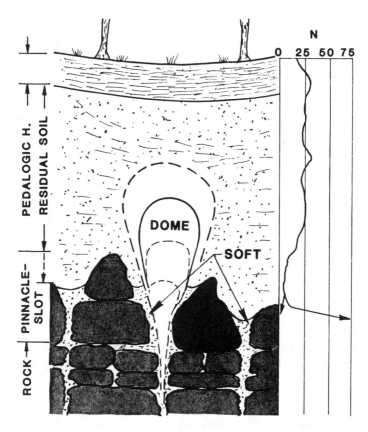

**LIMESTONE, DOLOMITE, MARBLE**

FIG. 2.10.   Residual Soil and Rock Profile in Solutioned Limestone with Significant Secondary Porosity in the Form of Vertical and Horizontal Cracks [The Relative Strength of the Residual Materials Are Indicated by the Standard Penetration Resistance, N, in Blows per foot (300 mm) of a 140-lb (63.6-kg) Hammer Falling 30 in. (762 mm) Required to Drive a 1.4-in.(35.6-mm) I.D, 2 in.(50.8 mm) O.D Sampler 1 ft (305 mm)]

concentrated in irregular layers or lenses in the residual soil; otherwise, it is randomly distributed.

The profile of soil strength as a function of depth is usually inverted compared to that of deposited soils. Deposited soil strengths tend to increase with increasing depth because of consolidation from the weight of the materials above. By way of contrast, residual soils over limestone are usually strongest in the uppermost 3 to 20 ft (1 to 6 m) and with more or less uniform strength for much of the remaining soil depth. However, beginning 5 to 15 ft (1.5 to 4 m) above the limestone surface, the soil becomes significantly softer with increasing depth. This is reflected in the *standard penetration resistance*, N, shown in Fig. 2.10. If the solution slots in the rock are wide with respect to the pinnacle widths, the pinnacles may punch upward into the firmer soil, impaled by the weight of the residual soil mass above. The firmer soils become wedged in the upper parts of the slots, with the softer materials confined to the deeper portions of the slot. This complex soil structure may be reflected in local settlement of the ground surface when the soil stresses are increased by construction, including the weight of structures supported at the ground surface.

The oldest soil is at the top of the residual soil profile; the youngest is at the bottom, which is the reverse of the related ages in deposited soils. The strength of clayey soils is largely the result of interparticle bonding forces that develop from molecular attraction plus the effective stress from soil weight. The intermolecular bonding increases with age. Both stress and age bonding increase with increasing depth in deposited clays because both the effective stress and age increase with increasing depth. In residual soils over limestone, the aging bonding decreases with depth. Moreover, the effective stress bonding may not increase with depth because the deeper younger soils may not have been stressed for a sufficiently long time to become fully consolidated.

Within the solutioned slots in the rock surface, the residual soil usually becomes rapidly softer with increasing depth; within the slots, the soft soil is shielded from the weight of the soil above. Immediately above a pinnacle, however, the consolidation stresses are increased. In this case, the soil may be locally somewhat stronger than the adjoining residual soil.

The boundary between rock and the residual soil above is abrupt: the product of solution weathering is an entirely different material structurally from the far harder unweathered limestone. This contrasts with other residual soils produced by the mechanical and chemical breakdown of the rock from the ground surface downward. In these residual soils, the weathering becomes progressively less with increasing depth and the strength and density increase with increasing depth.

The elevation of the boundary between limestone and its overlying residual soil is usually extremely irregular because of the solutioned slots and pits and cavities in the rock surface. In areas of advanced weathering, the blocks become pinnacles with wide trough-like slots between. When this develops, the term *depth to rock* is both confusing and very misleading. The engineering problem of supporting a heavy structure on such "bedrock" is both challenging and frustrating. Excavating to bedrock, as specified in many construction contracts, is an ambiguous requirement that spawns arguments, claims, and law suits.

Limestones and the residual soils derived from the limestone are sometimes covered with deposited soils of a variety of geologic origin: *tephra* (volcanic ash), *colluvium* (landslide and sloughed materials), and a wide variety of alluvial sedimentary materials. These hide the limestone and any residual soil. The weight of these materials increases the consolidation of the residual soil, thereby, improving its engineering qualities. But the deposited soils further obscure any solution-related defects in the underlying residual soil and rock.

# CHAPTER 3

# LIMESTONE SOLUTION AND ITS EFFECTS

As previously stated, the engineering problems in limestone terrain are largely the result of rock solution. Paradoxically, however, most of the engineering problems develop in the overlying residual or deposited soils that also mask those rock solution features. Thus, in order to solve the engineering problems, it is necessary to understand both the rock and any overlying soil, as well as their interaction to changes in the total environment. The major environmental factors are surface water infiltration (both natural and man-made), groundwater movement up and down with respect to the rock-soil interface, site excavation, filling, and structural loading.

## 3.1 CONTINUING DISSOLUTION OF PRIMARY POROSITY

Section 2.5 discussed the effects of the enlargement of the primary porosity by solution: decreasing the rock strength and increasing the compressibility. Continuing solution enlargement of the primary porosity also affects the rock mass geometry. Locally, the primary porosity coalesces to form large voids or *cavities*. The pores become elongated in the direction of groundwater percolation: vertically above the water table, and more or less horizontally below the groundwater level (with irregular deviations due to variations in rock composition and structure). These solution-enlarged cavities may join to form *caves* or *galleries*, usually deeper within the rock mass.

When the upper surface of the porous rock is above the water table, more or less circular pits often form in the rock surface. Some occur at random and others in regular patterns as if they had been man-made. Fig. 3.1a shows a cross section of such pits in an excavation for a footing foundation near Fort Lauderdale, Florida. Fig. 3.1b is an isometric drawing of the three-dimensional development of rock surface pits formed in limestones with significant primary porosity above the water table. In Florida and some of the Caribbean Islands, the pits are typically 8 in to 2 ft (200 to 600 mm) in diameter and from 1.5 ft to 10 ft (450 mm to 3 m) deep.

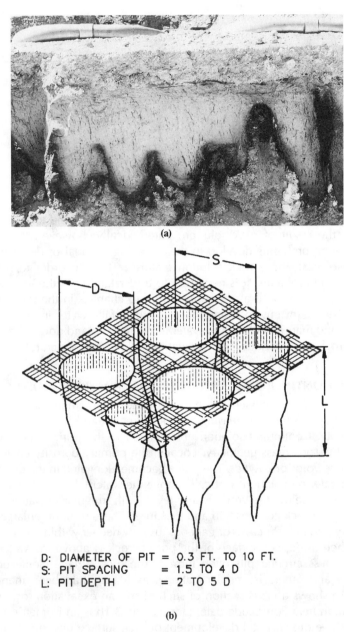

(a)

(b)

D: DIAMETER OF PIT = 0.3 FT. TO 10 FT.
S: PIT SPACING     = 1.5 TO 4 D
L: PIT DEPTH       = 2 TO 5 D

FIG. 3.1.  Solution Pitting of the Surface of Limestone with Significant Primary Porosity above the Groundwater Table: a. Cross-section of Pits 8 in. (200 mm) in Diameter and Filled with White Sand Exposed in a Foundation Excavation in South Florida; b. Isometric Drawing of Solution Pits

However, such pits underlying similar limestones in Tampa have been wider than 2 ft (600 mm) and as deep as 100 ft (30 m), based on pile foundation driving records. Pit spacings are varied, typically from 2 to 5 times the pit diameter.

Surface pits are unwelcome surprises in constructing foundations on the rock surface, and even more unwelcome when encountered in constructing deep foundations as will be discussed in Chapters 6 and 7.

### 3.2 SOLUTION-ENLARGED FISSURES: AGGRAVATED SECONDARY POROSITY

Solution of indurated rock with little or no primary porosity is also accompanied by rock disappearance: rock fissures enlarge and become slots that continue to become wider and deeper. The empirical evidence available (previously discussed) (Fig. 2.4) shows that the rate of rock removal by solution is not significant in terms of a human life span or in the functional life of most structures. (The rate of rock loss is typically between 2 and 4 mm per century in the Eastern U.S. where limestone solution problems are commonplace.) However, in geologic time, the amount of rock removed can be large.

As the rock fissures are enlarged into slots, the blocks defined by the fissures are correspondingly narrowed (Fig. 2.6). Eventually, the blocks may become slender spires, like the pinnacles in the "Stone Forest" in Kunming, China (Fig. 3.2), or loosely stacked blocks, like those piled up by a child.

Fissure enlargement at depth within the rock produces deep, narrow slot-like corridors or *cavities* with smaller openings along transverse fissures. Water circulation enlarges the fissures laterally into irregularly shaped tubes. Enlarged, elongated interconnected cavities known as *caverns* or *galleries* eventually develop. Although these are often natural wonders of great majesty and beauty (Fig. 2.8), they also can be hazards to human activity and structures on the ground surface above.

### 3.3 ROCK CAVITY ROOF COLLAPSE

When a rock cavity enlarges, the shear and tensile stresses in the cavity roof and the compressive stresses in the cavity walls increase, with the maximum shear stresses between, as shown in Fig. 3.3a. As the cavern continues to enlarge, tensile fissures may develop in the cavern roof, sometimes accompanied by diagonal shear fissures about midway between the cavern roof crown and the side walls. The tensile cracking allows blocks of rock to fall out of the roof, depending on the bedding and joint geometry of the roof (Fig. 3.3b). Many caverns have experienced some roof collapse in the geologic past, as demonstrated by occasional slabs of rock on the cavern floor that match the under side of the roof above. However, in terms of

(a)

(b)

*FIG. 3.2.    Stone Forest Near Kunming, South China [The Pinnacles Are 20 to 50 ft (6 to 15 m) High with Top Widths of 0.5 to 2 ft (150 to 600 mm)]: a. Irregular Pinnacles; b. Wider Pinnacles with Deep Fluting*

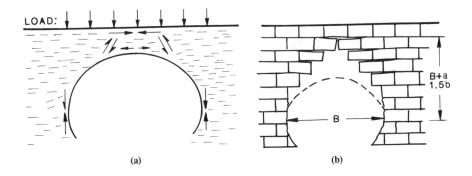

(a)                                                    (b)

*FIG. 3.3.    Stresses and Displacements in the Rock Surrounding a Cavity: a. Sound Continuous Rock with a Thin Cavern Roof; b. Displacements in Horizontally Stratified, Vertically Jointed Rock above a Cavern*

human life span, such roof collapses are very rare, as might be expected from the typical rates of rock solution (Fig. 2.4). Moreover, the roof of a cavern is not dissolved aggressively unless the cavern flows full. Thus, roof dissolution is slow in shallow caverns.

Local progressive upward roof collapse in thick but fractured rock strata is termed a *stope* (Fig. 3.4). This is a mining term for a narrow *roof fall* in closely fractured rock. The blocks forming the new roof above the fall are loosened by the fall. The result is successive rock falls and progressive loosening of the roof above. This creates a more or less vertical opening: the stope. (Stopes are often intentionally generated during mineral mining by excavating upward along a fractured mineral vein. The mineral ore falls into the mine tunnel below where it can be hauled out.) In a cavern, the process develops naturally; it can be a hazard to those persons who might be in the cave and a threat to structures on the ground above, if the stope propagates sufficiently upward.

The process ceases when the loosened rock that accumulates in the cave and the stope shaft fills the stope and prevents further loosening of the roof (Fig. 3.4b). This occurs because the volume of the collapsed fractured rock is greater than that of the intact rock by 1.5 to 2 times. If the cave is large enough or the accumulating debris is swept away by water flow, the stoping may slowly but indefinitely progress upward, ceasing when unfractured, intact rock is reached that can bridge over the opening. Otherwise it may continue upward to the rock surface. These vertical or near vertical stopes are termed *pits*. Pits sometimes connect caves at different levels in thick

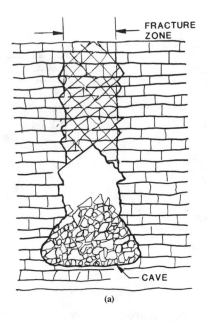

(a)

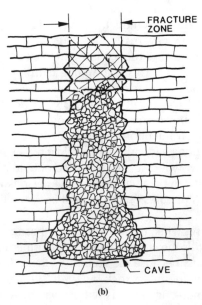

(b)

FIG. 3.4.    Stope Formation by Successive Rock Ravelling: a. Beginning of
Stope Development, Filling a Cave; b. Stope Debris Fills Stope Shaft, Stopping Further Ravelling

limestone formations. Pits are also formed by vertical dissolution at the intersection of steep, continuous fissures when there is significant downward water percolation.

When roof collapse or stoping intersects the rock surface, an open hole develops if there is no residual or deposited soil blanket above (Fig. 3.5). The opening is termed a *rock collapse sinkhole*, or simply a *sinkhole*. If the rock surface is blanketed by soil that is thin, the hole may propagate upward by successive shear failures in the soil. However, by far the more common mechanism of further propagation of the hole is ravelling and erosion of the soil into the hole generated by downward percolation of water. The resulting soil opening is also termed a sinkhole.

Although there are many examples of both forms of roof collapse sinkholes that have formed in the geologic past, the rate of formation of new rock collapse sinkholes appears to be extremely slow. The author has investigated hundreds of newly reported sinkholes; of these, perhaps two or three appear to be the direct result of rock collapse, including stope development. All three were in the Caribbean islands and none in the U.S. World-wide accounts of sinkhole development likewise indicate that new rock roof collapse and stope sinkholes are rare. The reason is that the rate of rock solution is so very slow. Instead, the usual common mechanism is the softening, ravelling, and erosion in the soil overburden that blankets the rock in much of the world areas underlain by limestone, as described in Sections 3.4, 3.5, 3.6, and 3.7.

Depending on the rock strength, fracture patterns, and the width of the stope or shallow cavern, the ground surface opening is usually initiated by

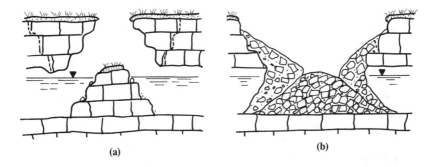

(a)

(b)

*FIG. 3.5.   Rock Collapse Sinkhole Above a Shallow Cavern in Horizontally Stratified and Vertically Jointed Rock: a. Initial Collapse with Rock Overhanging the Opening; b. Rim Collapse with Funnel-shaped Opening and a Debris Mound in the Cavern*

the fall of a truncated cone or pyramid of rock whose base may be nearly as wide as the cavity or pit below. The face of the remaining intact rock slopes away from the surface opening with increasing depth as shown in Fig. 3.5a. The average angle is about 45 deg in horizontally bedded rock with regularly spaced vertical joints: the angle of the maximum shear stress. However, if the rock joints are deeply developed, near-vertical and weathered, the falling mass may have vertical sides forming a rectangular or cylindrical hole.

With continuing ravelling of jointed rock in the sides of the newly-formed sloping hole, overhangs disappear and the hole eventually may become wider at the ground surface than at the cave level (Fig. 3.5b).

Although many areas underlain by limestone exhibit such rock collapse sinkholes, they are rare occurrences in terms of human life span. Rock dropouts only rarely damage structures or cause injury and loss of life because they occur so infrequently. However, the risk must be considered in design of structures above caves where the thickness of intact rock in the cave roof is less than about half the width of that cavern and when water flow conditions are favorable for continuing cavern enlargement. A recent rock collapse sinkhole in shallow limestone near a highway in the Dominican Republic is shown in Fig. 3.6a.

The cave below a rock collapse sinkhole initially is partially filled with blocks and fragments of rock and any residual soil that collapsed with the rock or subsequently sloughed in. Depending on the groundwater circulation in the cave and in adjacent cavities, the fine-grained materials, such as silt and sand, may be washed away. The larger limestone fragments also dissolve; however, the rate of solids movement by solution is a tiny fraction of the rate of erosion of soil and rock fines. The remaining rock dropout debris plus sloughing soil and organic debris from above may plug the open hole. This is followed by the accumulation of eroded soil and organic debris in the opening above the plug. In populated areas, trash may be added to the hole filling as shown in Fig. 3.6b. Such filling obscures the sinkhole. If the trash contains soluble materials, the filling may pollute the groundwater. Trash disposal in sinkholes is forbidden in most areas of the U.S.; however, this is often not adequately enforced. Eventually, because of trash fill and increased growth of weeds and shrubs, the hole may appear to be a small mound when viewed from a distance.

Water accumulating in the ground surface depression that accompanies most plugs in humid regions and water circulating in the cavity below may weaken the plug and cause it to subside periodically or even to wash out entirely. In one such plugged ancient (*fossil*) sinkhole in central Florida, the author found reasonably sound wood with a Carbon 14 dating of about 5,000 years, plus more recent animal bones and some contemporary trash,

(a)

(b)

*FIG. 3.6.   Rock Collapse Sinkholes: a. A New Rock Collapse Sinkhole and Two Older Sinkholes Filled with Water Near a Highway North of Santa Domingo, Dominican Republic; b. Trash, Debris, and Vegetation in an Old Rock Collapse Sinkhole in Tennessee*

irregularly mixed together at a depth of approximately 30 ft (9 m) below the ground surface. This suggests successive movements of the natural and man-made filling.

## 3.4 OVERBURDEN SUBSIDENCE FROM NEAR-SURFACE ROCK SOLUTION

In areas of closely spaced rock fractures, the rate of rock mass loss by solution increases because of the greater surface area of rock exposed to the percolating water. Eventually the rock mass near its upper surface resembles narrow ridges or columns, termed *pinnacles*, whose widths decrease toward their tops. The top of each pinnacle becomes lower as the rock is dissolved away. Where the rock fissures are closely spaced, the pinnacle tops become lower than those of the adjacent wider spaced pinnacles. The narrowing and the retreat of the pinnacle tops is accompanied by a sag in the residual or deposited soil blanket above (Fig. 3.7a). The author has termed such sags *solution depressions*. The narrower pinnacles may become impaled into the residual soil in the same way a wedge-shaped wood stake is impaled when driven into the ground (except the directions are reversed) (Fig. 3.7b). This also can cause a *solution depression*, particularly when the thickness of the overburden soil is less than the pinnacle spacing. If the overburden thickness is greater, the depression may not be noticeable.

The maximum depth of the ground surface depression is somewhat less than the loss of vertical thickness of the limestone stratum that was dissolved (although that thickness is usually difficult to estimate). The width of the subsiding area is about equal to that of the zone of greater pinnacle top depression, plus approximately half of the thickness of the soil overburden.

Solution depressions often become ponds in humid regions, accompa-nied by the accumulation of organic debris and sloughed and eroded soil from the depression rim. There is a difference in the solution depression accumulations compared to the materials that fill open sinkholes: the infilling of solution depressions is shallow, has a more uniform thickness, and is finer grained compared to that of the deep, cone-shaped dropout throat of a collapse sinkhole. Moreover, the fibrous organic material in collapse sink-holes tends to be thicker and less decomposed than that in the solution depressions.

Solution depressions also develop where there is a significant thickness of firm to stiff residual or deposited soil that can bridge over a small rock collapse sinkhole. The soil blanket sags slowly, bending and deforming plastically to form a depression in the ground surface, whose shape reflects that of the collapse opening. These depressions have the same appearance as the sags, due to the solution shortening of pinnacles. They can remain stable for centuries; however, some become open sinkholes by the process

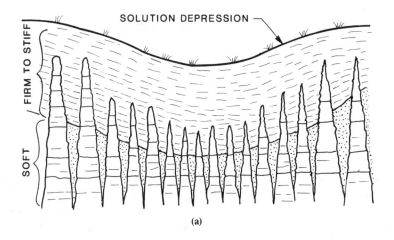

(a)

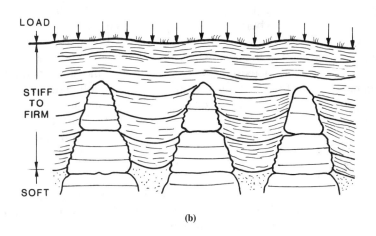

(b)

FIG. 3.7.   *Soil Overburden Subsidence from the Solution Enlargement of Joints in the Underlying Limestone: a. Effect of Variable Close Joint Spacing on the Level of the Pinnacle Tips; Close Spacing Causes Lower Tips; b. Pinnacles Impaled in the Stiff Overburden, with Soft Soil Filling the Deeper Portions of the Slots*

of soil stoping (*ravelling*) and erosion, followed by dome collapse, described in Section 3.6.

Areas of concentrated primary porosity also undergo aggravated near-surface dissolution. This may be followed by local crushing of the weakened limestone from the weight of any overburden above, accompanied by a sag in the overburden.

## 3.5 GROUND SURFACE SETTLEMENT FROM PINNACLE PUNCHING

When the vertical load on the overburden soil is substantially increased and the slots are wide and the pinnacles are narrow at their tops, the stiffer soil blanket becomes increasingly impaled on the pinnacles. This phenomenon is an aggravated stage of the local solution depression mechanism described in Section 3.4. As the pinnacles punch into the soil, the resistance increases in the same way a wedge gains resistance as it is driven into the ground. This is illustrated in Fig. 3.7b. The process is triggered by a substantial loading applied over a broad area, such as that produced by a high, wide embankment or a heavy building supported by spread footings or a mat foundation on the overburden. If the groundwater table is initially close to the ground surface and is then subsequently lowered significantly, the effective load of the overburden on the pinnacles is increased by the loss of buoyancy (an increase in vertical effective stress). This also generates pinnacle punching. Most of the aggravated punching occurs within a few days to weeks of the load increase. There is probably some small continuing subsidence, similar to secondary soil compression or soil creep.

The amount of the punching settlement can be estimated from a finite element analysis that computes both the stresses, strains, and elasto-plastic deformations in the overburden. This is described in Section 7.4.

## 3.6 CAROLINA BAYS

Consolidation and crushing of the solution-enlarged primary porosity of deep limestones buried under more than 100 ft (30 m) of overburden may explain some of the large, egg-shaped, shallow ground surface depressions in the Coastal Plain of North Carolina, South Carolina, Georgia, and north Florida, known as Carolina Bays (Fig. 3.8).

Borings by Law Engineering in the Southeastern U.S. have found that all of those depressions that have been investigated are underlain by a deep limestone strata with solution-enlarged primary porosity. The limestone has been found to be significantly thinner beneath the depression than the same stratum immediately outside the depression and the overburden thickness has been about the same whether measured in the depression or adjacent to it. Sometimes the limestone is absent or present only in the form of detached fragments or calcareous slimes. In other cases, the carbonates apparently have been leached away, leaving a skeleton that is predominately a loose cemented sand with little or no carbonate minerals remaining. Within the

*FIG. 3.8.   Carolina Bay in South Georgia, 1 mi (1.6 km) Long*

depressions, the strata underlying the solutioned limestone stratum are of uniform thickness, as are the non-calcareous strata above. There is no evidence of a sinkhole throat or channel to the ground surface. In the author's opinion, these bays are a special form of limestone solution depression peculiar to deep limestone strata that initially had significant primary porosity.

In the same area where the Carolina Bays have been identified, there are smaller solution depressions that resemble the Bays, except they are more rounded and without the clustering or orientation characteristic of the bays. These are often assumed to be filled sinkholes. However, borings in many of these depressions find the same thinning of a limestone stratum with well-developed primary porosity and no evidence of sinkhole throats. In the author's opinion, these are of the same origin as the Carolina Bays. The difference is that the associated limestone stratum is usually 50 to 150 ft below the existing ground surface. Experience has demonstrated that heavy structures, such as deep fills, can settle several inches in areas where the primary porosity of the deep limestone is highly solutioned. However, because of the thick overburden, little structural distress has been noted from this settlement. Although there have been no serious settlement problems or any real sinkholes developed in Carolina Bays, caution should be exercised in constructing settlement-sensitive structures in these areas.

## 3.7 SOIL RAVELLING-EROSION DOME FORMATION AND COLLAPSE

By far, the most widespread and serious limestone engineering problem, from the standpoint of frequency of occurrence and the hazard to life and property, is the development of a dome-shaped cavity in the overburden soil above a much smaller opening in the rock below. The sudden collapse of the roof of such domes is responsible for virtually all of the sinkhole failures that cause serious property damage and occasional loss of life world-wide every year.

The soil dome is generated by a combination of progressively upward slaking, ravelling, slab falls, and erosion of soil overburden into an open slot or hole in the rock below, as illustrated in Fig. 3.9. The rock opening need not be large; holes as small as 6 in. (150 mm) in diameter have generated domes more than 100 times as wide, depending on the strength and erodibility of the soil above the rock opening and the thickness of the soil overburden.

Water is the enabling medium, rising upward from fissures in the rock and seeping downward through fissures and the primary pores of the soil overburden. It causes sudden expansion and softening of dry or partially saturated clayey soils, termed *slaking*, on the soil-water interface. It causes surface weakening on soil-water, soil-air, and soil-rock interfaces. When cavities in the soil are present, thin fragments of soil separate from the mass and fall out, termed *ravelling*. This enlarges the cavities. Cavities in the soil become the focus of downward seepage because they shorten the path toward an outlet. If the seepage through the soil is significant, and the overburden thick and somewhat plastic, elongated sheets of soil, 0.5 to 2 in. (12 to 50 mm) thick and 5 to 10 times longer, sometimes peel away from the upper half of the cavity at the soil-air interface. This is sometimes termed *slabbing*. The soil fragments and slabs fall into the bottom of the soil cavity and accumulate in irregular mounds. The downward percolating water further softens and erodes the accumulating soil in the bottom of the cavity and transports the solids into deep slots or cavities in the rock surface. The process may stop temporarily when the rock opening is blocked by the accumulating slabs or by the viscous mud that develops. However, the accumulations eventually soften, if there is sufficient water, and they are carried into the rock openings below. Any flowing water in the cavity will eventually carry these materials away, leaving room for more sloughed and eroded material. Thus, the greater the water circulation in the rock cavities, the faster the process will proceed. Three conditions enable the ravelling, slabbing, and erosion processes to continue: 1) continuous or repeated wetting of the overburden soil, accompanied by 2) downward flow or percolation of groundwater into an opening in the rock surface, and 3) a hydraulic connection with water circulating in the cavities below or in a

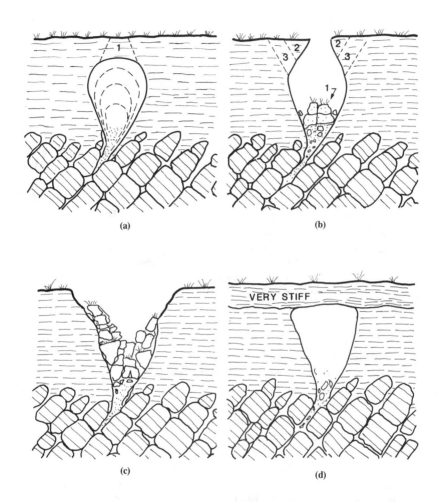

FIG. 3.9. Ravelling Erosion Dome Progression and Roof Collapse in a Cohesive Soil: a. Usual Inverted Tear-drop Shape with Potential Initial Dropout, 1; b. Initial Roof Dropout; Overhanging Rim and Successive Additional Dropouts, 2 and 3; c. Final Rim Collapse with Debris in Dome; d. Flat Dome Roof Beneath a Very Stiff Pedologic Horizon

sufficiently large cavity to continue to accept the solids that extrude or are carried in suspension into the rock openings.

Typically, the dome enlarges outward and upward toward the ground surface. If there is a point source of infiltration, such as a leaking pipe, the upward propagation of the dome will deviate toward that source. Sloping and even corkscrew dome shapes have been seen where the water source is not centered over a hole in the rock, or when there are sinuous, more easily eroded paths through the soil. The dome shape varies with the nature of the overburden: the more cohesive the soil, the wider the dome can become. Typically, the maximum dome width is from one-third to two-thirds (and occasionally more) of its height in clayey soils and less in soils of low plasticity and low cohesion. The dome usually is widest near its top, with the shape of an inverted tear drop (Fig. 3.9a) or more irregular sometimes resembling an upside-down pear. As the opening propagates upward, the slabbing, ravelling, and erosion are focused at its top and upper sides. The sloughed and eroded soil continues to accumulate near the bottom of the dome and intermittently squeezes, oozes, or flows into the rock cavities below depending on the rate of groundwater seepage downward. Usually, the rate of upward dome propagation decreases or stops when the dome approaches the much stiffer pedalogic horizons beginning at approximately 3 to 10 ft (1 to 3 m) from the ground surface. At that level, the top of the dome may flatten somewhat (Fig. 3.9d).

As long as the dome's soil roof is strong enough to arch over the opening, there may be no obvious reflection of the dome at the ground surface. Occasionally, a gentle sag develops from creep of the soil forming the roof of a very large dome. Sagging is often accompanied by the formation of circular tension fissures at the ground surface. The fissure circles are centered above the dome; their diameters are usually somewhat smaller than the dome's greatest diameter. The surface sag may collect ground surface runoff, thereby increasing water percolation downward. This often accelerates the dome's propagation upward. Sometimes near-vertical erosion holes a few inches (20 to 50 mm) in diameter develop on or near the circular tension fissures, resembling small animal burrows (rat holes).

When the roof of the dome becomes so thin or the dome so wide that the shear stresses in the soil roof exceed the soil's strength, a truncated cone of intact soil drops downward into the dome opening below, leaving a near-circular hole whose sides slope outward with increasing depth (Fig. 3.9b). Often, the top of the relatively intact truncated cone that dropped out can be seen in the opening below, with grass, shrubs, and even trees intact. (Be careful when approaching a newly formed soil collapse sinkhole; the overhanging wall may collapse under your weight and cause you to join the collapsed roof inside the soil dome.) With the passage of time, the overhanging walls of the hole slough and slide into the opening below, increasing the hole's ground surface diameter. Eventually, the hole becomes funnel-shaped, more or less centered over the rock hole below (Fig. 3.9c). Both the initial hole and the enlarged hole are termed a *soil ravelling sinkhole, or a soil collapse sinkhole.*

If the opening in the rock below the surface hole remains open, the surface opening may continue to enlarge by erosion and sloughing, sometimes to a diameter somewhat greater than the soil overburden thickness. If the opening below becomes clogged, sloughing soil and organic debris slowly fill the hole, often creating a sufficiently watertight bowl to hold water. A small pond forms that accumulates sloughed soil and decaying organic material, followed by peat, wood, and other natural debris (Fig. 3.10). The filling is often aggravated by using the hole for waste disposal, including large trash. Old dropouts can often be identified from a distance by the old automobiles and refrigerators and other large waste objects that may be difficult to dispose of otherwise.

The accumulating water in a plugged sinkhole sometimes erodes or displaces the soil plug, thereby draining the water. The filling process may recommence. If the groundwater should rise above the bottom of an open sinkhole, it can create a pond from below. The pond may also drain during periods of drought. One such ephemeral pond in Georgia included blind fish, apparently from a cavern system below.

Sinkholes are more prone to form in topographically low areas because the concentration of surface runoff provides greater infiltration potential. However, they can develop on slopes underlain by dipping rock formations. In areas of perched water tables, such sinkholes in slopes can become

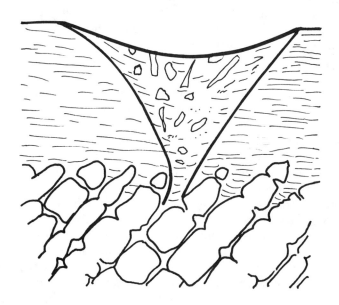

FIG. 3.10.   Sinkhole Infilling with Soil, Organic Material, and Debris

springs during wet weather. The discharging groundwater often erodes the down-hill rim of the sink, creating a well-developed (and sometimes disappearing) first- or second-order stream.

Sinkholes that form in cohesionless soils, such as those that blanket the porous limestones that underlie much of Florida, develop somewhat differently (Fig. 3.11). Cohesionless sand grains ravel and erode readily into small openings in the rock, similar to the flow of fine sand through the throat of an hour glass. When aided by downward water percolation, a rather narrow erosion channel or *pipe* develops and propagates more or less vertically upward. Eventually, a funnel-shaped sinkhole develops at the ground surface with its side slopes uniformly at the sand's angle of repose, typically 30 to 35 deg with the horizontal. This cone may continue to enlarge during rainy periods; however, the sinkhole can also enlarge during very dry weather when any temporary cohesion in the sand caused by capillary tension disappears.

### 3.8 "FOSSIL" SINKHOLE AND SOLUTION PITS

Sinkhole and solution pits that reach the ground surface are sometimes filled and obscured by soil and other deposition during later geologic events. Almost any form of deposition can be involved: Eolian (including volcanic ash), landslide and creep debris, marine, lacustrine, alluvial and glacial sediments, and sometimes lava flows. The sinkhole becomes partially filled and eventually covered up. If the deposition is thick enough, any topographic reflection of the opening is obliterated. The necessary thickness of non-conforming overburden to obliterate the sinkhole is 2 to 3 times the hole diameter, depending on the strength and rigidity of the new overburden.

In addition to concealing the holes in the rock surface, the non-conformable overburden changes the groundwater regime by developing new ground surface runoff and infiltration patterns. Moreover, the geologic events that caused the sinkhole filling nearly always drastically change the groundwater circulation in the rock that contributed to generating the original sinkhole and pits. The combination of events usually attenuates limestone solution in the sinkhole vicinity and any associated rock collapse.

Substantial changes in the groundwater regime in the old cavity systems below and in the ground surface infiltration can reactivate sinkhole activity by developing new ravelling-erosion domes in the unconforming overburden. These are preferentially developed above the fossil sinkhole where there is a existing path for erosion through the rock into cavities below.

In the author's opinion, the two largest sinkholes in Florida during the two decades between 1972 and 1992 were of this type. The first occurred in a dairy farm pasture in the south central part of the state in January of 1972 (Fig. 3.12). The soil overburden is a clayey sand lying unconformable above

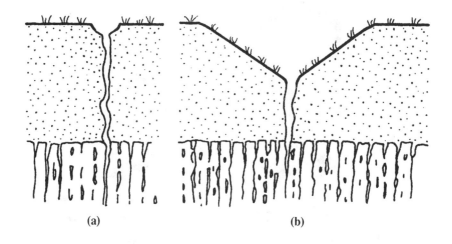

(a)                                          (b)

(c)

FIG. 3.11.   *Hour-glass Sinkhole in Overburden with Little or No Cohesion:
a. Early Development; b. Mature Sinkhole; c. Hour-glass Sinkhole North of
Bartow, FL*

(a)

(b)

*FIG. 3.12. Dairy Farm Sinkhole, Near Fort Meade, FL: a. Aerial View; Vehicle Tracks for Scale; b. Looking into the Cylindrical Rock Throat and Funnel-shaped Overburden Slope from the Sinkhole Rim*

a porous limestone. The failure took place without warning, exposing a cylindrical weathering-stained pit in the limestone surface 25 ft (8 m) in diameter and 50 ft (15 m) below the ground surface. Clear water was observed in the pit bottom. The initiating mechanism was probably large fluctuations in the groundwater level in the rock from pumping from nearby wells for mineral processing. It is not likely that aggravated surface infiltration was involved; the site was at the top of a low hill.

The second occurred in downtown Winter Park during the spring of 1981, and was briefly described in Section 1.1 and illustrated in Fig. 1.1. It severed two city streets, truncated a municipal swimming pool, swallowed a house, and damaged an automobile repair shop together with several automobiles parked behind it. The final hole diameter was more than 300 ft (90 m) with the cone-shaped opening approximately 50 ft (15 m) deep through clayey sands. Below was a 60 ft (18 m) diameter cylindrical opening in a soft, porous limestone. Although the geologic evidence is inconclusive, it appears that the opening represents an ancient pit or old collapse sinkhole in the limestone. Movement was apparently triggered by wide fluctuations in the ground level from well pumping and possibly aggravated by leakage from the swimming pool whose broken end can be seen in the right lower side of Fig. 1.1.

### 3.9 TOWER AND CONE KARST

Over long geologic time spans, continuing solution may cause many of the solution depressions and sinkholes to enlarge, combine and form deep, wide depressions surrounded by arc-shaped wedges or cone-like remnants of the rims. A new, lower ground surface level eventually develops, as the old sinkhole bottoms coalesce. Segments of the rims remain as fancifully shaped hills. Extreme examples of such exaggerated karst can be seen in warm humid regions such as Jamaica, Cuba, the Dominican Republic, and Puerto Rico in the West Indies and on the islands and mainland of South East Asia. The topography often is given interesting local names from objects they resemble: the *cockpits* of Jamaica, the *pepinos* and *mogotes* of Puerto Rico and the Dominican Republic (Fig. 3.13a), (pepino is a cucumber-shaped fruit), *las tetas* (teats) of the Philippines, the *cock spurs* of Thailand, and the *tower karst* and *cone karst* of Southeast China (Fig. 3.13b).

These towers and cones often contain caverns within them, possibly relics of higher groundwater tables in the geologic past. Suggestions of a new generation of karst developing in the low areas between the towers indicate the complexity of the groundwater table changes and ground subsidence.

FIG. 3.13.    Exaggerated Karst Land Forms: a. Pepino Karst in the Dominican Republic (The Same Term is used for Similar Topography in Puerto Rico); b. Tower or Cone Karst of Southeastern China

### 3.10  AGGRAVATING SINKHOLE ACTIVITY

Water movement is required to develop a sinkhole. Anything that increases the downward movement of water through the soil overburden can initiate or increase the activity. Experience demonstrates that three factors dominate: 1) increasing the infiltration of water at the ground surface,

2) depressing the groundwater piezometric level in the rock significantly below the soil-rock interface so as to provide an easy exit for the downward percolating water, and 3) repeated fluctuation of the groundwater from well above to well below the soil-rock interface that alternately saturates and drains the soil. Experience has demonstrated that both the increased frequency of new sinkhole formation and the development of unusually large sinkholes in the southeastern U.S. are related to extensive groundwater lowering, either by human activity or by drought. The sinkhole below the house in Fig. 1.1 was replicated several times in the same area of the small city over a period of several months, following a period of drought accompanied by increased groundwater pumping for water supply.

Mine and quarry dewatering throughout the world have been accompanied by increased sinkhole activity. The largest sinkhole in Alabama, and possibly the largest that has occurred in the United States during the past half century, developed as the result of quarry drainage. A broad valley about 30 mi (50 km) south of Birmingham, AL, is underlain by limestone formations of Ordovician age, approximately 450 million years old. They are blanketed by sandy and silty clay residual soils from 10 to 75 ft (3 to 23 m) thick. The valley topography is undulating with a few low swampy areas and adjacent low hills. The lower, more level areas are farmed; the more irregular high ground consists mostly of largely abandoned farms covered with pine forest.

One evening a retired farmer felt a slight shudder of his house, accompanied by barking and restless behavior of his dog. A strange sudden gust of wind occurred a few seconds later. The next morning the farmer walked back into the wooded area in the general direction of the sound and nearly fell into a hole which eventually became 300 ft (90 m) in diameter (Fig. 3.14). The rim was still sloughing, with slices of ground topped with trees sliding downward into the hole. The water level was 60 ft (18 m) below the ground surface based on later measurements. Soundings made a month later found the water 30 to 50 ft (9 to 15 m) deep. According to his neighbor, the man was so pious that his initial reaction was only, "Golly". The neighbor had a more colorful reaction!

Several intermittent curved fissures, approximately 0.25 in. (6 mm) wide, 5 to 15 ft apart, 30 to 50 ft (9 to 15 m) long and parallel to the sinkhole rim encircled the hole (the Golly hole). At irregular intervals, there were small vertical holes 2 to 3 in. (50 to 75 mm) in diameter along the fissure lines. The fissures were partially covered with ground litter, indicating that they had been present for months or possibly a year or two.

Aerial reconnaissance by the author showed that there was a large limestone quarry about 1.5 mile (2 km) from the sinkhole. The quarry extended more than 100 ft (30 m) below the groundwater table and continuous high pumping rates were necessary to keep the quarry dry. A few months before the sinkhole developed, the quarry floor had been lowered, which required an increase in the pumping rate.

*FIG. 3.14. The Golly Hole, 30 mi (50 km) South of Birmingham, AL: a. Aerial View Photographed approximately 1 month after the Collapse; b. Ground-surface View approximately 2 weeks after the Collapse*

The aerial reconnaissance also found two older, smaller sinkholes along a straight line extending from the quarry through the Golly hole: one was between the Golly hole and the quarry; the second was beyond the Golly hole. No persons could be found with knowledge of these older but newly discovered sinkholes that had been hidden by the dense pine forest. From their appearance, these were many years older. Both holes were approximately 50 ft (15 m) wide.

A very large sinkhole developed in 1962 in the deep mine area of South Africa, approximately 30 mi (50 km) from Johannesburg (Fig. 3.15). The site is underlain by approximately 300 ft (90 m) of cherty clay residual soil derived from dolomite. Gold-bearing rock is found at far greater depths. In order to mine the deep gold, it was necessary to pump water from the fractures in the gold-bearing rocks. The result was a depression of the groundwater table by approximately 400 ft.

Numerous sinkholes developed. There were so many that the mine owners employed a team of geologists and surveyors to map and evaluate these hazards for the safety of the mine, the miners, and the public. The deadly sinkhole occurred with no warning other than slow small settlement of an ore concentration plant. Possibly water leaking from the processing was a factor in the drop out. Suddenly, a hole nearly 300 ft (90 m) in diameter and more than 100 ft (30 m) deep swallowed the concentration plant and several small houses, burying 20 people. The soil overburden that collapsed into the hole covered the structural debris as well as the bodies, leaving no trace of either in the hole bottom, according to J. E. Jennings

*FIG. 3.15. A Large Sinkhole Near Johannesburg, South Africa, that Swallowed an Ore Processing Plant and Several Homes, Killing 29 Persons*

(1965). The vertical walls of the hole were still intact when the author made the photograph several years later. There was no evidence of rim collapse, probably because of the dry environment encouraged vertical cracking.

Similar but smaller dropouts have occurred above zinc mines in northern New Jersey near Bethlehem, PA. They have damaged houses and a highway but with no loss of life. As the rate of groundwater pumping decreased with the depletion of the ore, the rate of new sinkhole activity decreased.

The importance of increased surface water infiltration is illustrated by the sudden subsidence of one end of a large paper plant in eastern Tennessee. The structure was supported on shallow spread foundations on a stiff residual soil approximately 50 ft (15 m) thick over limestone. A large sinkhole was found immediately below a leaking drainage conduit that carried surplus process water back to the pumping station. The leak was not discovered for some time because the water disappeared downward. There was no hint of a problem until the dropout occurred. The remedial measures included rebuilding the water pipe and the building frame on deep foundations supported on the underlying rock. The repair cost in 1940 was approximately $1 million.

### 3.11 SUBSIDENCE AND SINKHOLE TERMINOLOGY

The terminology of dropouts and local subsidences is confused by the use of different terms for the same feature adopted from the different languages of those regions that have dropout problems. This book will use simple descriptive terms in English, because of the confusion in terminology that has developed. A number of references describe the different terms for the varied geomorphological features related to limestone solution. These should be consulted for detailed descriptions, as well as the hypotheses for the formation of these land forms. The **Encyclopedia of Geomorphology** (Fairbridge, 1968) and the **Glossary of Geology** (Gary, McAfee and Wolf, ed., 1972) are the most accessible, but list features alphabetically among numerous other land forms, making it difficult to attach a name to an observed feature. Texts in geomorphology describe the precesses but most do not emphasize karst. Three authoritative treatises devoted to karst are, **Karst Landforms** (M. M. Sweeting, 1972), **Karst Landforms** (J. N. Jennings, 1985), and **Karst Geomorphology and Hydrology** (Ford and Williams, 1989). The author has adapted the following more cryptic definitions from these three. Many of the terms are borrowed from other languages that do not convey to the non-expert reader what the words mean. Therefore, the author employs simple descriptive terms in English such as *sinkhole* and *solution depression* to help convey the nature of the features to those without specialized training in the geomorphology or the geology of karst terrains.

Dropouts with an open hole at the bottom and dropouts that have been plugged with impervious debris (usually with side slopes steeper than about 2 H to 1 V (26.6 de) with the horizontal) will be termed *sinkholes*. Surface depressions with no open hole in the bottom and with broad saucer or bowl-shaped bottoms will be termed *solution depressions*. The slavic word meaning valley, *doline*, is applied to both sinkhole and solution depressions by many karst specialists. However, because the depressions with the open holes usually are more hazardous and sometimes have a different mechanism of generation than those with more gentle slopes and no open hole, the author prefers their differentiation as stated above and does not use that ambiguous term.

## A Short List of Sinkhole Terminology

(Fairbridge 1968; J. N. Jennings 1985; and Sweeting 1972)

| | |
|---|---|
| Doline | A solution depression or a sinkhole. A circular or elliptical or irregular-shaped depression in the ground surface into which surface water from the surrounding ground drains. |
| Cutter | A solution-enlarged fissure or *slot* (A term borrowed from shallow mining of phosphate minerals in the southern U.S.) |
| Epikarst | The soft zone of unconsolidated residual soil immediately above the rock surface, and in the deeper slots between the solutioned pinnacles. |
| Grike | Solution-enlarged fissure or *slot* (used in U.K.) |
| Karren | Local surface dissolution or surface sculpture of limestone from running water, often in the form of parallel grooves. |
| Pinnacle | The narrow rock remaining between wide slots. |
| Polje | An extremely wide or valley like, flat bottomed solution depression, usually with an alluvial pavement in Yugoslavia and Turkey. |
| Pit | A more or less vertical shaft, usually a rounded stope or solution enlarged intersectioning fissure system. (See solution chimney and shaft.) |
| Slot | A solution-enlarged steep or vertical fissure. |
| Solution Canyon | A very wide and deep slot. |
| Solution Chimney | A solution-enlarged steep fissure or pit, sometimes somewhat crooked or twisted like a cork screw. |
| Solution Shaft | A near vertical more or less cylindrical hole from solution by falling water or stope activity. |
| Swallow Hole | A sinkhole in or adjacent to a stream into which the stream flow disappears. |
| Terrane | A variant of the word *terrain* used by geomorphologists and geologists to denote unity of geology, topography, and environment. |
| Uvala | A complex assortment of solution depressions or sinkhole with their bottoms a different levels and sometimes depressions within depressions. |

### 3.12 PSEUDO-KARST

Depressions in the ground surface resembling sinkholes and solution depressions over limestones can develop above formations other than limestones. Some of the susceptible formations are soluble, such as *halite* (rock salt) and *gypsum* (hydrated calcium sulfate) and its anhydrous form, *anhydrite*. In other formations in which the materials are not soluble, the subsidence begins with a cavity related to the origin of the rock, such as a lava cave or to man-made cavities, such as mines, tunnels, and sewers.

Gypsum, anhydrite, and halite are far more erodible than calcite and dolomite. However, these formations are not so widespread as the limestones. Moreover, they are often covered by such thick blankets of rock and soil overburden that there is limited groundwater movement and slow solution despite the high solubility. However, there are examples of karst features forming above such rocks. There are areas of karst in England developed above near-surface deposits of gypsum and anhydrite, as described by A. H. Cooper (1995). Cooper speculates that deflation rates have been observed as great as 1,000 mm per year compared to approximately 50 mm per 1,000 years or 0.05 mm per year for limestones in a similar environment as shown in Fig. 2.4a. The rate of deflation is 20,000 times greater.

According to Cooper, both solution depressions and ravelling, erosion sinkholes similar to those in carbonate karst have been identified in Yorkshire. Sinkholes 33 ft to 100 ft (10 to 30 m) in diameter and up to 66 ft (20 m) deep have occurred with major collapses every 3 years during this century. In addition, there are broad subsidence areas up to 2 km long and 100 m wide caused by solution of gypsum and anhydrite beds beneath thick strata of sandstone. He reports publications describing similar solution features in Germany and Poland.

Halite is so soluble that it is usually found buried deeply beneath the ground surface and insulated from significant groundwater movement, solution, and solution-related subsidence. However, in some regions where there are limited resources of salt, deep groundwater movement is generated artificially by solution mining of the salt. When there are sufficient fractures in the halite that water can circulate through when the hydraulic gradient is increased, the mining consists of recovering the salt water from some of the cavities by pumping from wells. If the natural groundwater circulation in the salt is not great enough, the circulation and solution rate are enhanced by pumping water into the salt through injection wells. In central New York state, severe pseudo-karst has produced individual sinkholes as large as 200 ft (60 m) in diameter and 60 ft (18 m) deep in the deposited soils above relatively impervious strata of shale and sandstone that overly the salt. These have disrupted a shallow perched water table, generating mud boils within the sinkholes where the bottom of the sinkhole has been depressed below the perched water table.

Downward percolation of water through shallow overburden above highly porous lava and fractured lava over lava caves can generate small sinkholes. Because the lava caves are small and the soil overburden usually is thin, the sinkholes are not as large or as widespread as those in soils above solutioned limestone.

Small sinkholes occasionally develop in residual soils produced by the in-place weathering of granite, gneiss, and some sandstones. These occur when the residual soils are easily erodible materials such as silty sands and sandy silts with little or no clay. It is necessary that the rocks be cracked, usually by folding, stress relief, or contraction upon cooling. The upper residual soils are eroded down into open fissures in the rock, developing small ravelling domes in the soil overburden and small irregular eroded channels in the residual soils in the rock fissures. Such features have been reported in South Africa, Australia, and observed by the author in the southeastern United States. These appear to be very localized and rarely of sufficient size to be serious hazards.

Sinkholes and subsidences resembling solution depressions occur over man-made underground openings such as leaking sewers (Fig. 3.16a) and poorly supported shallow mine shafts and adits (Fig. 3.16b). They also occur adjacent to deep excavations and quarries. The surface features mimic those that develop over limestones; only the destination of the eroded soil is different: man-made caverns.

Sinkholes are major hazards in cities with old sewer systems, because most sewers are beneath busy city streets. Moreover, municipal governments with limited resources often postpone sewer repairs until after trouble occurs. The surface dropouts usually occur without warning and so rapidly that they cause severe damage to both the pavement and to the vehicles travelling on them. A major factor in the generation of sinkholes over old sewers is that sewer flows often become much larger than had been anticipated at the time of the sewer design. Populations have increased since their construction. Moreover, storm water runoff rates are increased by large paved impervious areas in most cities. During large rain storms, the sewers sometimes flow under pressure, a condition for which they are seldom designed. Water exfiltrates from the sewer through deteriorating pipe joints or fissures in the sewer itself, saturating the ground. When the sewer flow returns to normal, the water in the soil seeps back into the sewer, eroding the surrounding soil and generating an erosion dome similar to those formed above solutioned limestone. The opening eventually collapses, usually during or immediately after an intense storm. Several cycles of exfiltration and erosion may be necessary to generate a dome large enough to cause surface collapse and the domes may be developing for months to years without any indica-

(a)

(b)

tion of their presence. Failure is typically triggered either by an unusually large storm or a series of smaller storms, or by some unusual surface loading such as a very large truck.

Mines cause sinkhole problems through both roof collapse and by ravelling and erosion from surface water infiltration into the mine openings. It is aggravated by the drainage that is necessary to continue mining, as discussed in Section 3.1.

### 3.13 EXAMPLE OF SUBSIDENCES AND FOUNDATION FAILURE FROM DOME COLLAPSE

An old airport west of Birmingham, AL, was seen to be an ideal site for light industrial and warehouse development because it was already level, free of trees, and bordered by a major highway and a railroad. Because the projected structures were largely one-story high and of simple construction, no foundation problems were anticipated, although the site was known to be underlain by limestone. A quarry nearby, adjacent to the railroad and highway, was being excavated in level-bedded sound limestone. The upper surface of the rock exposed in the quarry was seen to consist of pinnacles and slots 10 to 15 ft (3 to 5 m) deep in repeating rectangular patterns spaced 25 to 50 ft (7 to 15 m) apart. The pinnacles were covered with about 25 to 45 ft (10 to 14 m) of gravelly, sandy clay residual soil that was stiff immediately below the ground surface. Based on these observations, the developers of the warehouse site proceeded to construct roads and railroad siding tracks without further geotechnical investigation.

Although some of the railroad spur tracks to the warehouse development experienced some settlement despite no train traffic, there was no concern expressed by the project developers. Several one-story warehouses were built and occupied without problems. The structural engineer for one warehouse was more cautious. He noted the sags in the railroad track. Recalling past bad experiences in the area, he required a few soil borings before proceeding with design. Most of the borings found approximately 40 ft (12 m) of stiff clay over rock; however, one near the center of the proposed building found approximately 34 ft (10 m) of stiff clay underlain by 6 to 10 ft (2 to 3 m) of soft clay. The soft clay was below the level of rock found in the other borings. Still deeper there was no rock: only wet, pasty soil with fragments of rock. Because of this finding, coupled with his experience with other projects with similar soil profiles, the engineer recommended a pile foundation for the building. However, the architect convinced the proposed owner that piles were an extravagant waste of money. The owner, who was already upset because he had to pay for the borings, elected to save further expense by ordering that the building be supported on shallow spread footings on the stiff residual soil.

When construction was nearly complete, the column nearest the boring that found soft clay suddenly dropped into a hole approximately 20 ft (6 m) in diameter. The footing with part of the column intact could be seen in the hole bottom surrounded by fragments of the concrete floor. Looking into the hole, one could see that it became wider with increasing depth. The roof, still connected to the other columns, dropped, dragging the adjoining columns toward the hole, causing a portion of the outside wall to collapse. The cross section at the dropout is shown in Fig. 3.17.

The soil in the hole bottom was excavated to a level somewhat below the tops of pinnacles seen in the hole. Blasted rock from the nearby quarry, with pieces as large as 1/2 ton, were dumped between the pinnacles. Smaller crushed rock was placed above the coarse rock, developing a simple transition filter. Compacted soil was placed above the crushed rock, a new footing was built and the floor and roof were replaced. Nearly two decades later, the repair appeared stable. However, there is no assurance that another

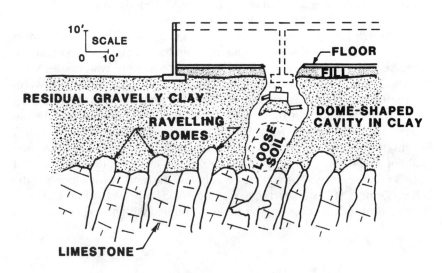

FIG. 3.17.   Sinkhole Beneath a Warehouse Building Near Birmingham, AL, that Swallowed a Footing Foundation and Precipitated a Roof Collapse

dropout will not occur at other locations (or possibly at that same location) in the future. Meanwhile, there have been several small subsidences under two of the nearby railroad sidings, and some minor settlement noted in other buildings in the development. The railroad spur tracks at two locations locally subsided as much as 1 ft (300 mm) from small circular depressions centered on the roadbed. It is likely that these were the result of developing sinkholes fed by water that accumulated in the crushed stone ballast of the roadbed.

All of the subsidences occurred along a straight path leading to the nearby quarry, which, by that time, was more than 100 ft (33 m) deep and 75 ft (23 m) below the water level in a nearby creek. At about the same time, the main railroad and a six-lane highway adjacent to the top of the quarry were both experiencing subsidences. A dropout of approximately 10 ft (3 m) in diameter occurred adjacent to the railroad. The railroad maintenance forces immediately filled the hole with broken stone. No further dropouts occurred, but local track subsidence continued. Train speeds were limited to 15 mph in that section to avoid derailing should there be further small track movements.

From time to time, both local settlement and 1- to 2-ft (300 to 600 mm) diameter dropouts developed in the adjoining highway pavement. These were repaired by filling the subsiding areas and the open holes with crushed stone and then repaving. After approximately 5 years of service, that section of the highway adjacent to the quarry consisted of patches on top of patches. Eventually, a reinforced concrete bridging trestle was constructed at ground level to support the highway. It rests on concrete-filled shafts drilled into the underlying limestone.

All the subsidences at the old airport, in the railroad, and in the highway commenced when the quarry depth reached approximately 50 ft (15 m), accompanied by high rates of pumping groundwater to keep the quarry dry. When the quarry pumping was stopped by court order (and impending damage claims), the subsidences and dropouts ceased.

This example is typical of many of the foundation engineering problems of structures constructed in areas underlain by solutioned limestone. A large proportion of the serious subsidences and concentrated occurrences of dropouts have accompanied or followed substantial lowering of the water table. Most are induced by human activity such as water supply wells and quarry and construction dewatering but some also occur naturally from protracted dry weather in normally humid regions. Virtually all have developed in the soil overburden: ravelling, erosion dome collapses. The initial dropouts exhibited openings that increased in diameter with increasing depth below the ground surface. Eventually, the overhanging edges collapsed, the soil sloughed, and the holes became widest at the top. Many appear to have been triggered by intense rain, leaking water and sewer pipes, water impoundments, or improperly channeled surface water. Borings in the sinkhole areas find a solutioned rock surface with pinnacles and with the slots between filled with soft wet soil. Cavities are found in the rock but there is virtually never any evidence of rock collapse contributing to the failure.

# CHAPTER 4

# GROUNDWATER IN LIMESTONE TERRAIN

### 4.1 SEEPAGE AND WATER FLOW RELATED TO THE SOIL-ROCK PROFILE

Chapters 2 and 3 discuss the nature of the materials that make up the soil-rock profiles in limestone terrains. Each soil and rock horizon exhibits a distinctive range in physical characteristics that influences the distribution and movement of water within it. These characteristics are principally the pore size, shape, interconnection, and the relative amount of pore space. This is expressed as either the percentage of pore space in the total rock volume, the porosity, or the ratio of the pore (or void) space to the volume of the solid materials, i.e., the void ratio as defined in Section 2.3.

The water and the solid materials interact: for example, water containing carbon dioxide in solution or with a pH less than 7 dissolves the solid carbonate minerals. Some solid materials retain water, either by molecular surface attraction *(adsorption)* or by capillarity in small pores *(absorption)*. Finally, the water pressure profoundly influences major engineering properties of the soils and rocks, particularly their strength, rigidity, and compressibility.

Characterizing the nature of water retention is sometimes difficult; characterizing the rate and direction of water movement is usually difficult and often impossible. A simplified diagram showing the relation between the soil and rock strata and the character of the water movements is shown in Fig. 4.1.

### 4.2 SATURATED FLOW OF WATER IN SOIL AND POROUS ROCK

Saturated flow in soil-like materials in the overburden and porous rock can be represented by Darcy's Law

$$Q = kiA \qquad (4.1)$$

| Cross Section | Horizon | Condition | Flow Regime |
|---|---|---|---|
| | Pedologic | Vert.Fissures | Laminar;intermittent,down |
| | Partially saturated residual soil | Nearly vertical fissures | Laminar; intermitten gravity down wet weather; capillary up; dry weather |
| | Saturated residual soil | Some vertical fissures | Laminar; mostly downward, some lateral flow |
| | Soft soil in deep slots | Soil interrupted by pinnacles | Laminar; downward, and occasionally upward |
| | Jointed Rock | Enlarged fissures network | Conduit and open channel, Turbulent flow; mostly lateral, but also upward and downward |

FIG. 4.1. *Relationship between Soil and Rock Strata and Groundwater Flow Regimes*

where $Q$ is the rate of water discharge from an element of soil with a cross section of $A$, perpendicular to the seepage direction. The energy gradient is $i$. It is a non-dimensional ratio of the energy loss along the flow path expressed as an elevation difference, $\Delta h$, to the length of the seepage path, $L$, described in ft/ft, or m/m. The resistance to seepage, which incorporates water viscosity, water density, and the size and shape of the seepage conduits in the soil or rock, is expressed by $k$, the *Darcy coefficient* or the *hydraulic conductivity* of the soil-water system. It has the dimensions of a velocity. In older U.S. publications, and in many originating outside the U.S., $k$ is termed the *permeability or permeability coefficient*. It has the dimensions of a velocity and is usually expressed in centimeters per second, feet per minute, feet or meters per day, and feet or meters per year. One cm per sec equals 2 ft per min, 873 m per day, 2880 ft per day, $1 \times 10^6$ ft per year, and $3.2 \times 10^5$ m per year.

The Darcy coefficient is the most variable of all the common engineering properties of a soil or rock. It can be as large as 100 ft per min (50 cm/sec) and as small as $2 \times 10^{-9}$ ft per min ($1 \times 10^{-9}$ cm per sec), a range of more than 11 orders of magnitude. It presumes that all seepage is by *laminar flow*, the orderly movement of the water particles in more or less parallel paths.

## 4.3 FLOW THROUGH ROCK WITH ENLARGED PORES OR FISSURES

Laminar flow breaks down when the pores are sufficiently large and the gradient sufficiently high so that the water particles move in chaotic paths. This is termed *turbulent flow*. It occurs in groundwater flow when the Darcy coefficient reaches the range of approximately 20 to 100 ft per min (10 to 50 cm per sec) depending on the hydraulic gradient and the void dimentions. This change over point in flow character occurs typically in materials such as gap-graded (*openwork*) coarse gravel, with open voids wider than approximately 0.25 in. (6 mm) and in limestones with the primary porosity enlarged by solution to similar sizes.

Flow through cracked, but otherwise non-porous, rock is largely laminar when the fissures are no wider than approximately 1/8 in. (3 mm) and the gradient less than approximately 0.01, for which Darcy's law applies according to Ford and Williams (1989, p. 142). Darcy's law also applies to limestones with primary pores smaller than approximately 0.25 in. (6 mm) in diameter, similar to medium gravel soils and gradients less than about 0.01. If the fissures and pores are larger and the gradients greater than approximately 0.01, the flow is likely to be turbulent and the head or energy loss occurs at a greater rate. A simple approximation of the seepage through such systems is to modify the Darcy equation on the basis that the velocity of turbulent flow in large conduits varies with the square root of the energy gradient, $i$, instead of being proportional to the gradient, $i$, as in laminar flow. The revised expression for seepage is

$$Q = k_{turb} A i^{0.5} \tag{4.2}$$

In this expression, $k_{turb}$ is the modified "Darcy" coefficient for turbulent flow. It expresses the roughness of the conduit walls, and the irregularities in conduit width and shape that control the turbulent flow head loss. Of course, if each conduit could be described in the terms required for analyzing turbulent flow in conduits or open channels, the diameter, wetted perimeter of the conduit, the roughness of the conduit surface, and the energy gradient (or conduit slope when the conduit is not full), a conventional computation of turbulent flow may be made as is described in texts on open channel and closed conduit flow. This is virtually impossible in groundwater flow because the parameters for the analysis cannot be readily measured.

## 4.4 SEEPAGE IN SOIL OVERBURDEN IN LIMESTONE TERRAIN

The soil overburden above limestone usually consists of three or more contrasting aquifers as shown in Fig. 4.1. These include: 1) the surficial pedologic horizons, 2) the firm to stiff residual or deposited soil strata that have been consolidated by overburden loads and desiccation, and 3) the soft, unconsolidated soft soil in the deeper parts of slots between the pinnacles and immediately above the rock, or in intermittently continuous thin lenses immediately above the rock surface. When there is a deposited soil above the residual soil, the system includes four or more potential aquifers.

The pedologic horizon is stratified: the upper foot or two, the topsoil and organic litter are more pervious, and the deeper zone of accumulation has low conductivity because of the accumulations of minerals from leaching of the topsoil above. When saturated by rainfall, the pedologic horizon becomes an aquifer for downward seepage. During dry weather, the degree of saturation decreases rapidly and any moisture is retained by capillary forces and there is no fluid flow.

The firm to stiff residual soil materials form a porous medium that is sometimes reasonably homogeneous. However, it also may consist of two layers of soil having the same grain size and plasticity but differing in that the upper part is only partially saturated during periods of high rainfall or when the piezometric level in the rock rises far above the rock-soil interface. The partially saturated upper part usually includes more or less vertical shrinkage fissures. The thickness of this upper part of the residual soil can sometimes be determined by test pits and large undisturbed samples that reveal the vertical fissures. Seepage through the saturated soils follows Darcy's law. There is little flow in the partially saturated portions: a combination of vapor transfer and seepage through capillary films.

The Darcy coefficient can be determined by laboratory tests of undisturbed samples or preferably by field tests using bore holes into which water is introduced into each stratum and the rate of water level fall by gravity or the rate of water inflow for a constant pressure can be measured. Such tests are briefly described in Chapter 5 and in texts on groundwater seepage.

Where vertical shrinkage fissures are present and the soil is saturated, the Darcy's coefficient is *anisotropic*: the vertical conductivity may be 10 to 100 times greater than that of the uncracked soil. The horizontal conductivity may be up to 10 times greater than that of the uncracked soil. Both laboratory and field tests for conductivity as well as seepage analyses must be made and interpreted considering anisotropy and non-homogeneity.

Flow nets and other seepage analyses can be used to model the seepage patterns below the groundwater table in the firm to stiff soils above the soft zone, with due allowance for vertical shrinkage fissures and any large lenses or strata of chert or sand-gravel inclusions. Methods for making such analyses are described in textbooks on geotechnical engineering and on groundwater seepage.

The bottom of the firm to stiff zone may be partially penetrated by the tops of the rock pinnacles. Although these locally may interfere with horizontal seepage, they seldom have a significant effect of the overall seepage in this stratum because the flow is usually vertical. The pinnacles reduce the area of flow but not the pattern of vertical water seepage. When there is a sufficient horizontal hydraulic gradient, the pinnacles may become small but seldom continuous dams that significantly interfere with horizontal seepage. Usually, only field measurements can determine the lateral seepage below the pinnacles.

The soft horizon immediately above the rock is usually very irregular in extent. It consists of the deeper parts of the solution slots in the rock that are filled with very soft clay. Those slots are discontinuously interrupted by the impervious rock pinnacles in all directions. The upper parts of the slots are usually filled with the stiff residual soils that have been forced into them by pinnacle punching. When the rock weathering has not advanced sufficiently to develop a slot-pinnacle profile, the soft horizon may consist of irregular, 1 to 3 ft (300 to 900 mm) thick partially discontinuous lenses above the rock and below the upper stiff residual soil horizon.

The pinnacles act as intermittent vertical barriers to horizontal seepage in the soft horizon. If there are openings in the rock bottom of the slots between the pinnacles that connect with rock cavities below, the soft clay may partially arch over the smaller ones. If the openings are sufficiently large, the soft clay slowly extrudes or flows into them, leaving small water-filled or air-filled voids in the bottom of the soft zone adjacent to the rock openings depending on the water table or piezometric level in the rock cavities. Often the rock exposed in an air-filled void is encrusted with partially dried clay.

Flow of water through the soft zone, when saturated, is likely to be laminar because of the high viscosity of the soft soil. The hydraulic conductivity is somewhat greater than that of firmer soils of the same classification. However, at high gradients, and with fluid pressures that exceed the minor principal stresses in the soft zone, the soft soil will *hydrofracture*. A plume of muddy water will displace the soft clay and will extend outward or downward, seeking an exit with a lower soil stress field, such as a fissure or cavity in the rock below. Where there are no exit fissures, the plume will enlarge laterally for great distances within the soft zone. Often, drilling water or drilling fluid is lost in the soft horizon when making exploratory borings or holes for remedial grouting.

The soft horizon is so erratic that meaningful modelling is usually difficult or impossible. At best, flow is anisotropic and non-homogeneous with scattered, small air-filled or water-filled cavities above the rock.

## 4.5 SEEPAGE IN LIMESTONE WITH SIGNIFICANT PRIMARY POROSITY

The high primary porosity of some completely indurated or consolidated limestones also means high hydraulic conductivity, $k$, if the diameter of the pores is less than approximately 3 mm as discussed previously. Without solution enlargement of the pores, the value of $k$ for some of the porous limestones in South Florida has been found to be from $2 \times 10^{-3}$ to 2 ft per min ($1 \times 10^{-3}$ cm per sec to 1 cm per sec), a range similar to that of poorly cemented sandstones to unconsolidated fine gravels. Although the hydraulic conductivities are somewhat variable or nonhomogeneous, flow can be approximated to be homogeneous over a sufficiently great distance that incorporates a representative range in rock porosity. The mass conductivity ranges from reasonably *isotropic* to *anisotropic*, with the hydraulic conductivity vertically sometimes as large as 10 times greater than the horizontal or as small as 1/10 the horizontal, depending on the geometry of the porosity.

The flow through such a mass can be modeled as that through a porous medium, with due allowance for anisotropy. The hydraulic conductivity can be estimated from a large number (i.e., dozens) of undisturbed samples that are large enough to include a representative range in porosities. A better alternative is field testing in which water is injected in the rock mass through a bore hole and the inflow rate is measured. The piezometric levels are measured in observation wells if the flow is unconfined, or by piezometers if it is confined by impervious strata. Injecting dye and tracing the apparent velocity of movement by the time interval between introducing the dye and its detection in observation wells can aid in determining the flow paths and the speed of the dye movement. (However, the velocity of the dye movement from the point of inflow to outflow is not the velocity computed by Q/A. To estimate the $k$ value of the rock, the dye velocity must be multiplied by the porosity of the rock.)

Where the rock pores have been enlarged by solution, $k$ (more properly $k_{turb}$) often exceeds 20 ft per min (10 cm per sec), as described in Section 4.3. Seepage is no longer laminar. However, head loss and flow can be approximated by using the modified Darcy expression previously discussed. The flow patterns in solutioned porous limestone are best determined by inflow tests, with water pressures measured by piezometers and the speed of water movement measured by dye testing. These are correlated with inflow measurements and sometimes with outflow measurements if most of the flow discharges through springs or into streams that can be gaged as described by Sowers (1981).

Engineers and contractors are often surprised by the high water flows through what has been depicted to them as "bedrock" limestone. This can lead to claims for extra compensation and expensive litigation in deep excavations below the groundwater level, as the following example illustrates.

A pumping station for storm water in Miami, FL, included a pit that measursed approximately 80 ft (24 m) square and 40 ft (12 m) deep, extending through limestone in which the primary porosity had been enlarged to the degree that it had the texture of petrified popcorn. The groundwater level was approximately 10 ft (3 m) below the ground surface. The construction specifications required placing all concrete foundations and walls in the pit under dry conditions. That requirement was interpreted by the resident engineer as being dry enough that he could strike a match on the rock surface. The contractor found that it was impossible to pump water fast enough to lower the water in the pit more than 10 ft (3 m), with the remaining 20 ft (6 m) still under water. Acting on the engineer's recommendation, the contractor tried grouting the rock in the sides and bottom of the excavation below the groundwater level. However, the grout washed away with the flow of groundwater into the excavation. Grouting was then repeated after allowing the water to refill the pit and thus stop the groundwater movement. This was more successful. However, it still was impossible to make the rock in the bottom "match-striking" dry. A geotechnical consultant retained by the contractor recommended that the excavation be allowed to flood. After flooding, the foundation slab would be poured using concrete tremied through the water. Eventually, the contractor received permission to do this. The concrete quality proved to be excellent and the foundation and side walls were completed by pouring concrete under water with no further trouble. This demonstrated that this method should have been employed in the first place. The contractor sued for damages, claiming faulty specifications and bad decisions by the resident engineer. He won both direct and punitive damages because of the unreasonable interpretations of the specifications and arbitrary decisions of the design engineer and his resident engineer.

## 4.6 FLOW IN ROCK FISSURES AND CAVITIES

The flow in the rock fissures is more complex. Without solution enlargement, the fissures may exhibit either laminar or turbulent flow, depending on their width and the energy gradients discussed previously in Sections 4.2 and 4.3. Two rock flow conditions occur: *open channel* and *conduit*. The open channel flow involves only partial filling of the rock cavities; the piezometric level is below the cavity roof and varies with time and location depending on the rate of water flow. This is shown in Fig. 4.2a except for the lower-left channel. The conduit flow is analogous to that in interconnected water pipes; the piezometric level is always at or above the conduit roof. This is shown in Fig. 4.2c. Fig. 4.2b shows both conditions. Rational analysis of a

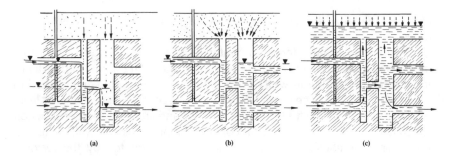

FIG. 4.2.    Idealization of Storage, Inflow, and Piezometric Levels in Inter-
connected Rock Cavities and Soil Overburden: a. Low Rainfall and Little
Inflow from Upstream; Multiple Piezometric Levels in the Rock; b. Moderate
Rainfall and Significant Flow from Upstream. c. High Rainfall and Large
Inflow from Upstream so that the Piezometric Level in the Rock Cavities
Rises into the Soil Overburden

conduit system requires specific data on the size shape and elevation of each
conduit and on the conduit interconnections. At best, reliable information
approximately the size and shape of limestone cavities is fragmented. Therefore, analysis must evaluate a range of possible conduit and water storage
systems that fit the available measured data on gradients, dimensions, and
discharge. The results must be viewed as possible representations of the
actual flow regimen. Even if the results of an analysis can be verified by field
piezometric and flow measurements, such analyses should be viewed as
parametric studies because all the conduits and their interconnections cannot be identified.

### 4.7 PIEZOMETRIC LEVELS—WATER TABLES

Because of the contrasting aquifers (i.e., the firm to stiff overburden, the
irregular soft soil immediately above the rock, and the complex fissures and
solution channels in the limestone), several different piezometric levels can
exist simultaneously at one site location. These can include multiple artesian
and perched aquifers.

The water pressures and piezometric levels in the limestone on a site
depend largely on 1) the relative inflow to and outflow from the fissures and
cavities on the site through rock cavities adjoining the site and 2) the storage
of water in the cavities on the site. (The site as used in this chapter is a
geographically small area ranging from a few acres or hectares to hundreds
of acres or hectares that is being considered for a particular project). Inflow

from downward percolation of water through the pervious soil overburden is very slow and usually does significantly influence the water pressure in the rock on the site except during long periods of unusually great or prolonged rainfall or when there are significant man-made sources of water in the soil overburden.

The rate of inflow to the site is controlled by the groundwater levels, the groundwater storage, and the hydraulic conductivity upstream of the site. (*Upstream* means that the hydraulic energy or piezometric level is greater than that on the site; therefore the groundwater flows in the downstream direction.) The rate of outflow from the cavities on site toward the rock downstream is controlled by the groundwater levels, the conductivity of the cavities, and the storage downstream. It also may be controlled by conduit obstructions at the downstream side of the site that acts as weirs or orifices that limit the groundwater flow outward. The difference between the inflow into the sit, downward percolation on the site, and outflow from the site is the amount of water that either is stored in the cavities on site or released from those cavities on the site.

Depending on the interconnection of cavities and fissures with the sources of water inflow and the outflow path, there may be more than one piezometric level in the different limestone cavities on site, as shown in Fig. 4.2. This complicates and confuses the groundwater flow and pressure picture, particularly when only a few piezometers or observation wells have been installed to measure water levels on the site. A significant number of observation wells or piezometers at several levels and at several locations is necessary to identify multiple water pressure systems.

When the inflow, outflow, and storage combine to develop a piezometric level in the rock that is lower than the bottom of the overburden (Fig. 4.3a), water in the soil overburden seeps slowly downward. However, the free water surface falls very slowly because of the low hydraulic conductivity of the usual residual clayey soils. The water is *perched* in the overburden and is independent of the piezometric level in the rock. The groundwater pressure in the soil is close to atmospheric, because of the energy loss caused by the downward seepage. With a well-distributed hydraulic connection between the soil overburden and the rock cavities below, the downward seeping water, plus any loose fine-grained soil in the soil fissures, slowly joins the water in the cavities.

The water in the soil voids or pores above the piezometric level in the overburden is retained by capillary tension, sometimes termed *soil suction*. The *capillary fringe* is the upper limit of capillary saturation and is equivalent to the capillary tension in the larger voids in the capillary fringe. Above the top of the fringe, the soil water is discontinuous: retained in the smaller voids. At the level of capillary saturation, the soil moisture evaporates (but at a very low rate). Any soluble solids in the water precipitate at this level,

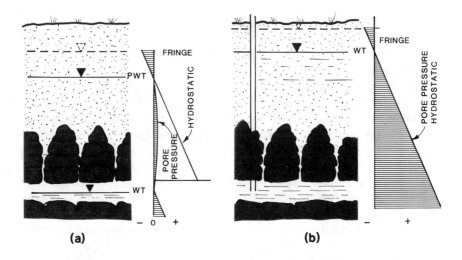

FIG. 4.3.    Effect of Groundwater Tables on the Groundwater Movement in the Soil Overburden: a. Piezometric Level is within the Rock Below the Base of the Soil Overburden. Groundwater in the Overburden is Perched, with Slow Downward Percolation; b. High Piezometric Level in the Rock Generates a Continuous Groundwater Pressure System in both the Rock Cavities and in the Overburden. Downward Seepage Rate Reduces, Depending on the Piezometric Changes

forming a relatively impervious layer that sometimes partially cements the soil. The cemented layer is termed a *hardpan*. It greatly reduces the hydraulic conductivity and greatly increases the soil strength at that level. Above the capillary zone (and any hardpan), the soil is only partially saturated; any water movement is slow and in the vapor phase during dry weather.

During rainfall or snow melt, the piezometric level in the overburden rises. Downward seepage continues at approximately the same rate because the additional head is absorbed by the head loss in the longer seepage path.

Surface water percolation may soften the upper portions of the firm to stiff soil overburden if the soil had been hardened by desiccation. Thus, if a sufficiently large erosion dome has already developed in the soil so that its roof is close to the ground surface, softening of the upper soil may cause the roof of an existing dome to collapse. Many sinkholes develop during the later stages of heavy rains or immediately afterward, particularly when the rain follows a long period of drought and low groundwater levels.

When inflow to the limestone cavities on site exceeds the outflow, water is stored in the cavities and the piezometric level in the rock increases, sometimes rising far above the base of the soil overburden (Fig. 4.3b) and merging with the groundwater in the soil. When the rock cavity piezometric level rises above that in the soil overburden, there is a short period of upward seepage in the soil overburden until equilibrium is reached and the two groundwater systems merge. At that time, the groundwater in the combined system becomes hydrostatic and vertical groundwater seepage in the soil stops. The groundwater pressure in the soil below the water table rises from virtually zero to hydrostatic. This may have profound effects on the soil. Water now fills any voids or fissures that were partially air-filled. This may weaken the adjoining soil, making it more susceptible to erosion and structural collapse. During a period of heavy rain and infiltration, the water table in the soil rises, causing downward seepage. The seepage stops when the two systems reach equilibrium again. When the water pressure in the rock falls, seepage resumes downward. Eventually, equilibrium may be reached with two water tables, as shown in Fig. 4.3a.

During periods of rapid and large fluctuations in the piezometric levels in the rock, short periods of artesian pressure in the rock but perched water tables in the soil overburden may develop. The alternating upward and downward seepage and the accompanying pressure increases and decreases in the soil voids aggravate soil erosion in the overburden. This condition is particularly conducive to enlargement of soil fissures and of erosion, ravelling domes.

## 4.8 RELATION OF GROUNDWATER AND CLIMATE TO SINKHOLES

The discussions in Chapter 3 show that sinkholes are generated by downward seepage through the overburden that weakens the soil overburden and erodes the soil into the fissures and cavities in the rock. There are two conditions that are conducive to this erosion: 1) lowering the groundwater table from considerably above the rock-soil boundary to below it and 2) increasing the downward percolation by increasing the water available at the ground surface. Lowering the water table increases the gradients in the vicinity of openings in the rock surface and promotes soil erosion. In addition, the effective stress in the soil increases as the buoyant effect of the water is lost, which leads to local soil failures and the tendency for soil to break loose at soil water interfaces. Repeated fluctuation of the water level from above the rock-soil interface to below it particularly aggravates sinkhole development.

Increased water at the ground surface also increases the ravelling-slabbing-erosion process. This is particularly important when the soil overburden has high hydraulic conductivity. Prolonged periods of precipitation and snow melt are typical natural conditions leading to sinkhole development. Human sources include leaking sewer and water pipes, faulty ground surface drainage, intentional disposal of storm water into pits excavated in the ground surface, and surface water impoundment without impervious lining in the reservoir. All aggravate the downward seepage by substantially increasing the head difference and the downward hydraulic gradient.

One of the worst examples of the effects of increasing water at the ground surface occurred near Albany, GA, in July, 1994, following large area flooding generated by a tropical storm as reported by Hyatt and Jacobs (1995). The residual soil and an irregular stratum of deposited soil are very sandy silty clays with greater hydraulic conductivities than more typical limestone overburden. The flood water depths were as high as 20 ft (6.5 m). More than 300 new sinkholes were identified and measured. Three-fourths occurred in an area of approximately 5 sq mi (14 sq km) that were inundated; the remainder occurred outside but close to the inundated area. The largest was elliptical, approximately 145 ft by 69 ft (44 by 21 m); the median was circular, approximately 6 ft (1.8 m) in diameter.

Four conditions are unfavorable to ravelling, soil erosion, and dome collapse. First is a relatively constant piezometric level that is well above the base of the firm to stiff overburden. This means a small, downward hydraulic gradient and less erosion. Second is a rock cavity piezometric level that is well below the rock surface, accompanied by little or no surface infiltration or other sources of downward percolation that might initiate ravelling and erosion into openings in the rock. Third is a stratum of erosion-resistant, low-conductivity sandstone, claystone, or shale immediately above the limestone that resists water percolation into the limestone below. Fourth is extensive impervious areas at or near the ground surface, such as paved parking areas or wide impervious embankments (with surface drainage that leads off site) that shield the overburden soil from infiltration and thus inhibit downward percolation. However, the surface drains that remove the runoff from the impervious areas must not leak and counteract the ground surface shielding. Draining surface water off the site can generate sinkhole activity elsewhere. This must be considered in the overall site impact on the environment.

There are few published studies that relate the groundwater regime (including surface infiltration) to either ravelling and erosion dome formation or sinkhole development. The Florida Sinkhole Institute (W. L. Wilson 1988) published the results of a study of a two county area in the vicinity of Orlando, FL. During the 26-year study period, 129 new sinkholes were reported. (The reporting of new sinkholes, however, was not uniform. During the first 20 years, which relied on voluntary uncoordinated reports, only 54 were identified. During the last 6 years, when a systematic effort was made to obtain and document the reports, 75 were identified. The average number per year for the uncoordinated reporting was 2.7; the number per year for the period that was systematically documented was 12.5, almost 5 times greater.)

The study found two significant trends. First, by far the greatest incidence of new sinkhole development occurred at locations where downward percolation of water was expected to be greatest, such as in topographic lows, particularly those with standing water during dry weather. Second, the greatest rate of sinkhole development occurred during periods in which the groundwater level in the underlying limestone was the lowest: 23% of the sinkholes occurred in the 2% of the time that the groundwater level was very low. By way of contrast, 18% of the sinkholes occurred during the 13% of the time when groundwater was the highest. These data apply to a limited area of Florida where the sinkhole development is unusual because of the high primary porosity of the limestone, the variable nature of the residual soil, and the surficial deposits of pervious sands. However, these observations are compatible with the author's experience elsewhere in diverse geologic formations and environments.

The repeated wide fluctuation in the rock piezometric levels, from well below the rock surface to well above the rock surface and into the upper overburden, is probably the most favorable regimen for aggravating ravelling and erosion dome development, based on the author's experience in the southeastern United States. The rising water pressure level saturates the soil overburden and the subsequent falling level aggravates downward seepage, ravelling, and erosion. A second condition favorable to dome formation is a low piezometric level in the rock coupled with some unusual source of water in the overburden. However, unless the soil has high hydraulic conductivity, contains vertical fissures that concentrate seepage, or is highly erodible (sometimes termed dispersive) clay, the rate of dome development is likely to be slower than that which is generated by widely fluctuating piezometer levels in the limestone that saturate the overburden.

### 4.9 EXAMPLES OF SINKHOLES INDUCED BY HUMAN ACTIVITY

Much of the increased sinkhole activity in the southeastern United States during the past three decades is related to the changes in the groundwater levels and infiltration caused by human activity. Lowering the groundwater level drastically, often accompanied by cycles of lowering and recovering, appears to be the most important cause of aggravated sinkhole activity. In Florida and south Georgia, with increasing population and industrial activity, the groundwater levels have been depressed locally as much as 50 ft (15 m) by pumping wells for water supply—individual, municipal, and industrial. In addition, lowering the water level during strip mining for phosphate minerals and to secure water for the processing of the ore also have lowered the groundwater. Some counties in central Florida have experienced tens of new sinkholes per year. Some sinkholes have been extraordinarily large, such as the 300 ft (90 m) diameter sinkhole in Winter Park (Fig. 1.3) or groups of smaller ones such as that which swallowed a home in Bartow (Fig. 1.1).

Even the victims of these human-induced sinkholes have learned to live with them. Other persons have a dual relationship with them, as illustrated by the owner of a fern farm in central Florida. The owner's home was on one side of the farm, approximately 100 ft (30 m) from his well that supplied domestic water and irrigation water for the ferns. In addition, during very dry weather, irrigation water was supplied to the trees that provide shade for the ferns.

During a prolonged drought, the pumping rate was unusually large. One morning, the family awakened to find a hole that measured approximately 60 ft (18 m) in diameter and approximately 10 ft (3 m) deep within a few feet of the house. During the morning, the hole slowly became deeper, until by noon it was nearly 60 ft (18 m) deep. Suddenly, the sandy clay soil in the bottom of the hole began to boil, with the appearance of cooking oatmeal. The boiling was followed by a sound like a giant toilet flushing. Water quickly rose to within about 8 ft (2.5 m) of the ground surface as seen in Fig. 4.4. The farm owner notified the county officials who placed a fence

FIG. 4.4. *Sinkhole Generated by Local Lowering of the Water Table by Wells to Provide Irrigation Water for an Adjoining Fern Farm*

around the hole and ordered the owner to vacate the house. (The author could not find out if the well pumping was discontinued). Of course, the well was essential for the farm, the income of the farmer, and eventually the income of the county, so it is likely that some pumping continued. The pump was operating during the author's site visit, several days after the failure). Fortunately, the farmer had sinkhole insurance, which paid off. Within a few days of the insurance payment, the farmer and his family reoccupied the house. As far as the author knows, the well and the farm continue to operate as before the failure. This illustrates the dilemma of living in sinkhole-prone terrain.

The impact of an increase in water in the soil overburden is illustrated by the settlement of a paper manufacturing plant in Tennessee, an area underlain by deep, solutioned limestone. The plant foundations were spread footings supported on the stiff residual soil overburden 50 to 60 ft (15 to 18 m) thick over the rock. Approximately 15 years after the plant went into operation, one end of the large manufacturing building began to settle into a cone-shaped depression in the residual soil. A water pipe beneath the plant floor had been leaking, possibly for more than a year. It generated a typical ravelling, erosion dome that suddenly collapsed. The plant production came to a halt until the pipe could be repaired and the foundations underpinned. Manufacturing resumed, but with more careful accounting for the water utilized in manufacturing. The cost of repair was more than $1 million almost 50 years ago. It would be several times that today.

The experience of a small college in northwestern Georgia also illustrates the effect of shallow impoundment as well as depressing the groundwater level. The college owned extensive land, much of which was used for farming in which the students participated to help earn their tuition. In recent years, the value of the farm in producing income decreased and additional sources were explored. Much of the center of the site was an old flood plain with the groundwater table within 5 to 10 ft of the ground surface and with a small creek winding through it. The area was underlain at depths of 10 to 20 ft (3 to 6 m) by hard limestone, which when crushed made a high-quality concrete aggregate. A local quarry proposed that the college lease them an area at one end of the property approximately 2500 ft (450 m) from the college buildings for an aggregate quarry.

With increasing demand for concrete aggregate for housing, road construction, and industry, the quarry flourished and enlarged. The quarry royalty payments became a substantial part of the college income.

As the quarry was excavated deeper, finding even better rock, it was necessary to pump large quantities of water out of the ground to make it possible to continue to operate. Suddenly, sinkholes began to appear in the fields adjoining the quarry. At first, this seemed to be no serious problem, because the quarry had stockpiles of sub-standard rock (quarry spoil) that needed disposal. The spoil was regularly used to fill the sinkholes as they developed in order to prevent the cattle from falling in. However, the problem became serious when the new sinkholes approached the college buildings a half mile away along a more or less straight path. That path followed the same direction (*strike*) as the rock joint fissures in the quarry. The college faced a serious financial dilemma: Was the value of the royalties sufficient to justify the risk of the new sinkhole formation? Eventually a decision was made to continue the quarry, but to limit dewatering to the immediate area being quarried.

Meanwhile, the college enlarged a small lake on the campus. It had been developed as a memorial to students who lost their lives in military action beginning with World War II. A few months after the lake was enlarged, it suddenly went dry. Two sinkholes developed in the center of the enlarged area (Fig. 4.5). These were repaired with the waste rock or quarry spoil dumped in the sinkhole throats followed by clay compacted above it. The lake was refilled, only to disappear 2 days later with both the initial sinkholes plus some new smaller ones. Eventually, the lake addition was filled with clay fill on top of quarry spoil, blanketing the sinkholes and restoring the lake to its original size.

In all these examples, the sinkhole activity probably could have been avoided (or at least postponed) by avoiding the activity: the fern farming, the paper manufacturing, the quarry, and the lake enlargement. However, with the increasing scarcity of both land and water, that is not always a viable alternative. Instead, the potential for sinkhole generation must be considered in land-use planning in areas underlain by limestones and measures must be taken to minimize groundwater changes that aggravate sinkhole development. Structures whose failure could cause loss of life should be designed to be insensitive to sinkhole activity and the risk of sinkhole development should be made known to the public that occupies such areas. As will be discussed in the subsequent chapters, measures can be taken to avoid the most sinkhole-susceptible areas. Measures can be taken to prevent sinkhole activity and to correct it when it does occur. The economics of the preventive and corrective work, as well as the impact on the environment, must be included in the decisions regarding if, where, and when to build.

*FIG. 4.5.   One of a Pair of Sinkholes that Opened Simultaneously in the Bottom of the Newly Enlarged Section of an Ornamental Lake on the Campus of a Small College in Northwest Georgia*

# CHAPTER 5

# SITE INVESTIGATION

## 5.1 OBJECTIVES

The objectives of a site investigation for geotechnical design and construction in limestone terrain are the same as for similar projects in other soil and rock formations. First, determine if any potential geologic hazards are present, such as landslides, earthquakes, faulting, volcanic activity, flooding, and, specifically, sinkholes. Second, determine the depth, thickness, and potential engineering behavior of the soil and rock under the site that will affect the proposed project during its expected life. Third, determine the depth and fluctuations of the groundwater beneath the site.

Because of the peculiar nature of limestone terrain, these objectives are intensified where the mechanisms of limestone solution and the defects produced by those processes are concerned. The investigation identifies the degree of dissolution and the pattern and extent of specific hazards, such as sinkholes, soil ravelling, and erosion domes, and the potential for their further development. More detailed information includes the length, height, width, and orientation of the solution-enlarged fissures in the rock and the slot and pinnacle geometry.

The thickness and strength profile of the soil overburden, and particularly the thickness of any soft zone over rock, influence the design of shallow foundations. The location of rock collapses and the potential for soil dome collapse also impact the use of shallow foundations. Adding fill on top of the rock or the soil overburden, as well as excavating soil overburden and rock, change the present and future integrity of the system because of the changes in stress in the underlying formations, changes in infiltration, and interference with surface water flow. These must be evaluated. Understanding the potential for environmental changes that effect solution and rock or soil dome collapse is necessary in estimating the risks involved with long-term use of the site.

The existing groundwater conditions, the potential changes that occur naturally, and those that occur from human activity, are evaluated so that steps can be taken to minimize adverse effects. Finally, the potential risks in utilizing the site must be understood by all concerned: the owners, the workers on site, the public utilizing the site and affected by conditions on the site, the public officials that administer land use, and the general public.

## 5.2 RECOGNIZING SINKHOLE AND SOLUTION DEPRESSIONS: KARST TERRAIN

Areas underlain by solutioned limestone exhibit characteristic topographic features that identify karst terrain as well as pinpoint the locations that are most likely to become engineering problems related to the rock solution features. The most obvious features that can become problems are the existing sinkholes and solution depressions. If their depths exceed the contour interval on published topographic maps, they appear as closed contours, properly hachured on the inside to denote a depression. Depressions shallower than the contour interval may not appear on topographic maps. These can be identified from stereoscopic examination of overlapping air photographs in the same or adjoining flight path. Even a casual viewing of individual aerial photographs (Fig. 5.1a) shows many of the depressions. If the ground surface is relatively free of trees and large shrubs, more subtle sinkhole and solution depressions can be identified from ground surface reconnaissance and from low-level aerial reconnaissance (Fig. 5.1b). Aerial viewing early in the morning or late in the afternoon is helpful because even shallow depressions are accented by shadows. If the depressions are water-filled, reflections of sunlight with a low sun angle also accent those depressions (Fig. 5.1c).

*False color infrared* air photography emphasizes the more vigorous vegetation growth and accompanying greater infrared florescence by displaying it as bright red on the photo image. (False color infrared film displays growing green vegetation as red; green color, such as paint, as blue; water as blue, even if it is brown or black; and bare soil as gray or yellow.) In dry regions with low water tables and pervious soils, the moisture effect on vegetation is sometimes reversed. The depression bottom can be the driest spot with little or no vegetation because water can percolate more readily into the rock below. The false color infrared image of a dry solution depression is pale pink to yellow and gray.

Sinkholes, and even the deeper solution depressions, sometimes become so filled with trash, thick shrubs, or small trees, that they do not appear as topographic lows when viewed from the air, on aerial photographs, or from a distance on the ground surface. Instead, they may be identified by the jumbled appearance of trash or the mound-like appearance of thick vegeta-

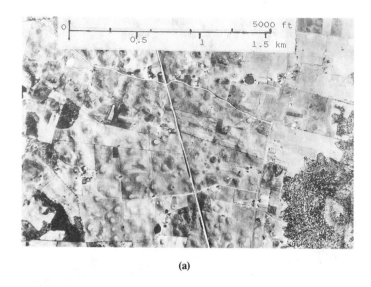

(a)

(b)

*FIG. 5.1. Aerial Photographs of Limestone Terrain in Northwestern Kentucky; a. Vertical Aerial Photograph of Sinkhole Topography at a Scale of 1:20,000, (on the Original) for Mapping Purposes; b. Low-Level Oblique Photograph Showing Solution Depressions, Ponds, and Sinkholes from Height of 1,000 ft (300 m) above the Ground Surface*

tion (Fig 5.2). These anomalies are easily spotted in crop lands or pastures where the holes and depressions are avoided in cultivation. In wooded areas, they may be difficult to identify because of the trash and thick

FIG. 5.2.    Trash Hidden by Shrubs, Making a Sinkhole Appear as a Mound
from the Distance [Photograph taken During Winter When the Leaves were
Absent (in Western Kentucky)]

vegetation, even on the ground surface, until one stumbles into them, as has
the author on occasion. Clumps of trees and dense shrubs in otherwise
cleared or cultivated ground in limestone terrain should be considered
suspect, unless demonstrated otherwise by close inspection.

Karst terrain is also characterized by the absence of the first and second
orders of surface water drainage. If the soil overburden is relatively impervious, the system of *first-order drainage* (i.e., shallow erosion gullies and
small wet-weather streams one can step across) is poorly developed. If water
flows in the first-order drainage, the water often disappears by percolating
through a more pervious overburden zone or through small holes in the soil
termed *swallets*. These drain into fissures slots in the limestone below
instead of joining with other first-order drainage to form *second-order* streams
(i.e., those one must jump or wade easily across). Absence of or very wide
spacing of first-order, second-order, and sometimes *third-order streams* (this
one must wade across) is characteristic of solutioned limestone in humid
regions.

Larger streams, third-order and greater, that enter limestone terrain
sometimes disappear into *swallow holes* that connect with rock cavities
below. They may reappear as springs or seeps further downstream. These

reappearing flows may disappear during dry weather and reappear during wet weather. Disappearing streams usually reappear when their path crosses into non-limestone terrain or in nearby river valleys as springs. Dye tracing is usually necessary to establish the relation between the disappearing stream and its reappearance elsewhere.

Stream gaging at several locations along a well-developed second- or greater order stream underlain by solutioned limestone often finds decreasing rates of flow instead of increasing flow proceeding downstream. If the stream crosses a lithologic boundary, beyond which cavities are small or absent, the flow often reappears as springs in the stream bed and banks.

Patterns of sinkhole and solution depression locations usually reflect patterns of structure in the underlying limestone. Sinkhole and solution depressions often occur along straight or slightly curved lines (termed *lineaments*) above prominent joints or other structural weaknesses in the underlying limestone. Several lineaments can be seen in Fig. 5.1a. The larger sinkholes are often found at the intersections of lineaments of sinkhole and solution depressions and a greater risk of new sinkholes is present at lineament intersections.

### 5.3 PRELIMINARY INVESTIGATION

Preliminary investigation begins with an intensive review of existing information. As in other site investigations, the fundamental data include geologic maps and reports of the area to identify the underlying rock and, particularly, any limestone formations and their approximate geographic boundaries. The mapped boundaries are seldom precise, particularly in limestone terrain, where the rock is usually hidden by a blanket of residual soils. The name of the formation, its geologic age, and its structure are keys to correlating experience with previously-identified limestone solution features in the region with the potential for developing solution at a particular site. U.S. Geological Survey (USGS) topographic maps may reveal specific clues to solutioned limestone: 1) surface depressions with no outlet, 2) the absence of first- and second-order streams and the wide spacing of third-order streams, 3) springs and disappearing streams in humid regions, and 4) lineaments of streams, small ponds, and depressions.

In areas of significant cave development, local, regional, and national cave exploration groups have prepared and compiled descriptions, detailed maps, and reports on specific caves. Information may be obtained from the *Cave Research Foundation* (117 Hamilton Valley Road, Cave City, KY) and the *National Speleological Society* (Cave Avenue, Huntsville, AL), as well as from local speleological clubs. While such information may not be available for a specific site, any information about local caves is useful for correlating cave occurrences with the limestone formations at the site and with any

topographic similarities. Experiences with subsidences and collapse affect-
ing structures are often recorded at local offices of the State or Province
Department of Transportation, the County Engineer, the County Environmental Officer, the City Engineer, and local geotechnical engineering and
geological consultants. In some areas where collapse sinkholes occur frequently, the local, state, and federal agencies involved with emergency
management of disasters from storms and earthquakes, such as FEMA in the
U.S., sometimes develop detailed maps of sinkhole occurrences.

The second tool of preliminary investigation is some form of *remote
sensing*: examining the site from the air or from space for the topographic
and geomorphic features that are typical of solutioned limestone and the
closed depressions, sinkhole, and the absence of low-order streams in humid
regions. Usually, the most valuable tool is air photography (black-and-white,
color, and false color infrared), with sufficient overlap in adjacent photographs so that stereoscopic study is possible. For most of the United States,
such photographs are available from the U.S. Geological Survey, the U.S.
Forest Service, and the U.S. Department of Agriculture. The Forest Service
and Department of Agriculture photographs are made at more or less regular
time intervals of approximately 6 years; the Geological Survey photographs
are at longer intervals, as needed for map preparation. Although the scales
of the photographs, typically from 1:15,000 to 1:40,000, are too small to
reveal small, but important, limestone solution features, they are very useful
resources. Their ready availability, low cost compared with new (but more
detailed photography), and the periodic re-photographing, which may show
if new solution depressions or sinkholes are developing or old ones are
changing with the passage of time, make them a unique tool. In areas that
are presently occupied by industry or obscured by urban sprawl, the earlier
photos, made before the land was disturbed by construction, are more likely
to reveal solution features.

If new photographs are justified by the size and cost of the project,
larger-scale photographs, 1:2500 to 1:10,000, reveal more detail. Color
photographs, while having slightly poorer resolution and somewhat higher
cost than black-and-white photographs, often reveal the presence of depressions, springs, and streams by subtle color differences. For example, a
solution depression may retain moisture during dry weather. It will exhibit
more vigorous vegetation growth accompanied by stronger greens in color
photography than the surrounding water-deficient vegetation. Similar differences can be seen more clearly in the infrared florescence, depicted as bright
red in false color infrared photography. Conversely, an area above a large
soil erosion dome may exhibit weaker green in color and only faint pink in
false color infrared because the soil holds less moisture and the vegetation
is stunted. Areas underlain by closely spaced rock fractures that drain
moisture from the surface soil sometimes exhibit yellowish greens or weak

pink images. Areas immediately above shallow rock pinnacles or blocks also may exhibit poor vegetative growth because of the limited depth of soil, if the vegetation is normally deep rooted, such as alfalfa, as is illustrated in Fig. 5.3.

When air photographs are not available, or are of inappropriately large scale, and there is insufficient time for new photographs, reconnaissance, by either fixed-wing or helicopter aircraft, is useful. In many cases, the direct reconnaissance has the advantage of looking at the site from different oblique angles in order to examine a cliff face or the overhang in a fresh sinkhole. The helicopter has the advantage of slower speed and the ability to hover, but the disadvantage of longer flying time to a site from an airport and much more severe vibration, which makes high-resolution photography difficult.

The views can be recorded by hand-held photography, such as Fig. 5.1b and 5.3, using high-resolution black-and-white, color, and color-infrared film. All are available in 35 mm and 120/220 (62 mm) widths.

FIG. 5.3. *Alfalfa Field in Central Tennessee Shown in a Low-level Oblique Photograph [the Light-toned Areas are Stunted Alfalfa; the Darker Areas are Tall Vigorous Alfalfa; the Stunted Patches Reflect the Lack of Soil Moisture Caused by Sound Limestone at a Depth of between 0.5 and 1.5 ft (150 and 450 mm); in Deeper Soil, the Roots Extend More Than 3 ft (1 m) Deep]*

Satellite imagery is useful in identifying large features such as lineaments of solution depressions and similar topographic features. The resolution of images available to the public is usually insufficient for features such as sinkholes that are smaller than approximately 50 ft (15 m) in diameter and springs and first-order streams that are often hidden by vegetation. However, the images can be helpful in viewing the total picture of a region.

## 5.4 ON-SITE RECONNAISSANCE

On-site reconnaissance begins in the preliminary project planning stage and should continue through construction. This includes examining the area for verification of changes in previously observed or photographed features. This also includes suspicious terrain details that are difficult to see from the air because of tree and vegetation cover as well as overhangs and other obstructions. With good air photo interpretation, the time required for ground surface examination can be reduced and focused on suspicious areas. This is a major tool in the preliminary investigations when permission for site access is often limited or impossible.

## 5.5 DIRECT INVESTIGATION

Direct investigation includes a number of procedures that require site access and varying degrees of invasion and physical disruption of the ground. The same boring and sampling techniques used for foundation exploration in other formations are applicable, as listed in Table 5.1.

Because of the irregular pinnacle and slot topography of the rock surface and the cavities in the rock, the number of borings required per unit of site area in limestones is usually much greater than for site investigations in other formations. If the rock formations are horizontally bedded, it is likely that most of the joint fissures will be vertical. This makes it very difficult to identify the fissures and their spacing and widths with vertical borings. Angle borings are required to find steep or vertical fissures and to define their width and filling. Regardless of the dip of the rock, a few angle borings are usually desirable to identify randomly oriented fissures and solution features.

The fallacy of using only vertical borings in formations that could include very steeply dipping strata or fractures is illustrated by the failure to find vertical solution-enlarged joints beneath the partially completed foundations of a nuclear reactor in central Tennessee. A very closely spaced grid of vertical boring was used to explore strata of level bedded limestone, containing widely spaced shale seams. The site investigation included several hundred vertical borings whose purpose was to define small variations in the nature of the limestone using downhole geophysical probes (Section 5.5), as well as to identify any significant caverns. No angle borings were

## TABLE 5.1.    Direct Investigation Technique

- Percussion drilled holes to identify the soil-rock boundary. The observed drilling rate is an indicator of rock hardness and rock discontinuities such as fissures and voids. This requires recording the rate of penetration for short intervals of drilling: minutes per foot or per meter, as well as visual examination of the drill cuttings. However, it is difficult to differentiate between a large boulder, a pinnacle, and the upper surface of continuous rock.
- Test borings with intact split-spoon samples and Standard Penetration Resistance Tests in soil-like materials, particularly in the soft zone immediately above rock and the soil in cavities within the rock after drilling into the rock. The bore holes are made by augers or rotary cutters using air or drilling fluid to remove the cuttings. Any loss of drilling fluid is measured as an indication of the size and continuity of rock fissures and cavities. Laboratory tests of the samples provide data for accurate classification of the soils and for estimating some of the engineering characteristics such as hydraulic conductivity and response to load.
- Undisturbed sampling of the stiff overburden at representative intervals and, if possible, of the soft soil overlying the rock and filling slots and cavities. The size of the sample tube is determined by the boring diameter. Laboratory tests of the samples provide quantitative data for engineering analyses of the soil hydraulic conductivity and response to loading.
- Cone penetration tests in soil, particularly in the soft soil zone and in soil in the rock (after core drilling or percussion drilling exposes the soft seams). The usual cone point and sleeve resistance can be supplemented by pore water pressure sensors using a *piezocone*.
- Core boring preferably with triple tube diamond bits in rock. When the rock is so weak or closely fractured that the core recovery is less than approximately 90%, larger diameter cores, 4 to 6 in. (100 to 150 mm), are preferred.
- Oriented core drilling to determine the dip and dip azimuth of the strata and of fissures.
- Large diameter drill holes that permit human access to examine the exposed materials directly, particularly at the soil rock interface. The minimum diameter is approximately 30 in. (760 mm); holes 36 in. (900 mm) or larger are preferred. Direct access requires casing in the hole for safety, which means alternating drilling, setting casing, and making the observations. This is often impractical below the groundwater level.
- Test pits or test trenches to the rock surface, exposing both slots and pinnacles. This requires either flat slopes or bracing to prevent cave-ins and to provide safety. It is often impractical below the groundwater level despite heroic pumping to unwater the bottom. Moreover, pumping could trigger sinkhole activity. Pits and trenches make it possible to view the stratification, the orientation of fissures, as well as the geometry of the soil-rock interface in three dimensions.
- Borehole photography or video imaging of the borehole walls.

made because the large, high-speed, truck-mounted drills available at the site were limited to near vertical holes.

Excavation for a cooling water channel adjacent to the reactor-structure found several joints that had been enlarged by solution into narrow caves, 2 to 4 ft (0.3 to 0.6 m) wide and 4 to 7 ft (1.5 to 2 m) high. These can be seen in Fig. 5.4, exposed in the bottom of a cooling water channel excavated

*FIG. 5.4.   Two Solution-enlarged Vertical Joints in a Horizontally Stratified
Limestone Between Two Shale Strata (the Joints Intersect Behind the Face of
a Deep Channel Excavated to the Lower Shale Stratum, as Depicted in the
Photograph)*

40 ft (12 m) deep in the limestone. Two of the caves intersected at right
angles, forming a small chamber under the nearly completed reactor foundation.

The initial response of the regulatory authority was to order removal of
the completed 8-ft (2.5-m) thick concrete mat foundation, followed by
excavation to the cave level, approximately 34 ft (10 m) below the foundation level. The objective was to uncover any other caves that might be
present and to replace all the excavation with concrete.

After more detailed study, including entry into the caves and mapping
them, it was found that only a few such enlarged vertical joints were present.
These caves and larger joints were cleaned out by a combination of manual
mining and pressure washing, and then filled with concrete. This delayed
construction of the reactor foundation, on the project critical path, by several
months, with a total cost of more than $1 million. The alternative of
removing the reactor foundation would have cost approximately 10 times
more. Angle borings, oriented to identify and explore the solution enlargement of the vertical joints, would have disclosed the problem. Ironically, the
geologic examination of an exposure of the same limestone nearby disclosed

a similar rectilinear pattern of near vertical joints, most of which exhibited dissolution. This exposure could have warned the investigators of the need for angle borings to investigate the nature of the deeper portions of the joints.

Borehole photography and its successor, video imaging, provide a picture of the wall of an uncased borehole. While the photographic camera makes intermittent images, the video camera provides a continuous view of the hole as it is lowered or withdrawn (so long as the hole walls remain stable). A photograph of a video image of fractures in the wall of a 4-in. (100-mm) boring in limestone is seen in Fig. 5.5. Each of the two successive

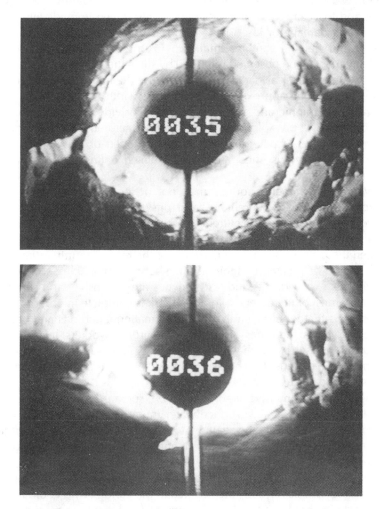

FIG. 5.5.   Photographs of a Continuous Video Image of Fractures in the Wall of a 4-in. (100-mm) Diameter Borehole in Limestone

images depicts approximately 1 ft (300 mm) of hole centered about the depth indicator (in feet) in the image center. Fractures appear as irregular dark areas in the circular band of light. The vertical black line is the support for the downhole light. The continuous video image resembles the view a person would have riding in the front of a moving subway train; the walls are a circular band of light with a black spot (the camera) in the middle. Both the photographic and video cameras can be used below dirty groundwater by flushing the hole with clear water before and even during viewing. If the materials to be viewed are prone to caving, a casing is necessary to stabilize the hole walls. The hole is advanced 1 or 2 ft (300 to 600 mm) ahead of the casing and the camera or video camera is introduced immediately (but at some risk of damage or loss). The video record for such alternating casing advance and viewing contains small gaps, 1 or 2 in. (25- to 50-mm) wide. The detail that can be observed is limited by silt in the water below the water table. Fissures appear as dark shadows because the light does not penetrate fissures further than approximately twice their width.

The photographic and, particularly, the color video examination of the borehole makes it possible to evaluate closely fractured rock that produces no core recovery, to determine if fissures are open or clay-filled, and to determine if the voids encountered are isolated enlarged joints or large continuous caverns.

Even large diameter borings may not define the extent of slot and pinnacle formation. Test trenches and pits to the rock surface may be required. The size and depth of trenches and pits is limited by the equipment available, the strength of the overburden soil, the extent of the soft zone between the pinnacles, and the hardness of the rock. The minimum trench length should be greater than the distance between pinnacles, so that a full slot width is exposed. Pits and test trenches are expensive, and should be made after most of the other exploration work is complete in order to define the locations that would gain the most information. Such pits and trenches can be hazardous: measures are necessary to protect workers, livestock, and the curious passersby from harm. When the rock surface is below the piezometric level, large boreholes, trenches, and pits may be impractical or impossible, no matter how desirable.

All holes from invasive site investigation, such as borings and test pits, should be protected against surface water inflow. Casing extending above the ground surface should be sealed into the ground surface in all observation wells to prevent surface water from entering them. Open holes and pits are avenues for downward seepage for surface water that can aggravate erosion at the rock surface and possibly precipitate sinkhole development and collapse. Moreover, they are hazards to the curious who could fall into them. Test pits should be filled with compacted soil that is no more pervious that the surrounding soil. The borings and wells should be filled with sand cement grout after they have served their purposes.

## 5.6 GEOPHYSICAL EXPLORATION

### 5.6.1 The Geophysical Approach

Geophysical exploration involves measuring a force system in the earth and inferring boundaries between zones of similar response to those forces by the patterns of force change known as *anomalies*. The engineering behavior of the materials within each zone sometimes can be deduced from attenuation or distortion of the forces compared with what would be expected in a homogeneous material or by the velocity of propagation of the force wave. The force systems can be natural, such as gravity and the earth's magnetic field, or induced, such as an electrical current or a sound or shock wave. The methods are indirect in that, while the force patterns measured are related to the boundaries and engineering properties of the ground, they do not measure the boundaries, and most do not directly provide the data on physical properties needed for computing ground stability or for foundation design.

The methods that have proved useful in soil and rock exploration in limestone terrains are listed in Table 5.2. The notations of where the measurement is made have the following meaning: GS = made at or near the ground surface, BH = made at depth in a borehole.

TABLE 5.2.   Geophysical Exploration

| Natural Forces | Where Measured |
| --- | --- |
| Gravity | GS |
| Magnetic Field | GS |
| Self Potential | BH |
| Natural Gamma Ray Emission | BH |
| Induced Forces | |
| Ground Penetrating Radar | GS, BH |
| Seismic Refraction | GS, BH |
| Seismic Reflection | GS, BH |
| Direct seismic wave transmission | |
|  Up-hole, down-hole seismic | BH to GS, BH to GS |
|  Cross-hole seismic | BH to BH |
| Sonic Profiling | BH |
| Electrical Resistivity | GS, BH |
| Induced Electromagnetic Field | GS |
| Gamma Ray Attenuation | BH |

The ground surface tests usually involve making measurements at many different locations, either along lines (*traverses*) or in a grid pattern. The results are interpreted from the variation of the force with the location on the site, and, in some cases, with the spacing between the sensing points. The ability of surface methods to identify solution features can be expressed by the *anomaly ratio, $R_a$*:

$$R_a = B / Z \qquad\qquad (5.1)$$

where $B$ = the anomaly width and $Z$ = the depth from the ground surface (or the distance from the sensor to the anomaly). The wider the anomaly compared to its depth, the larger the ratio and the more likely that it will be detected. Combinations of downhole and surface geophysical measurements make it possible to increase the depth at which a defect can be identified. *Borehole geophysical probe* tests introduce groups of instruments (the probe) in a borehole as discussed in Section 5.7. These make it possible to test soils and rock at great depths below the ground surface, but typically at only small distances from the borehole.

### 5.6.2 Ground Surface Natural Force Fields

Gravity and magnetic surveys detect anomalies in those natural force fields by instruments at the earth's surface. Gravity reflects differences in soil and rock density; magnetism reflects minerals such as magnetite that distort the magnetic field. Gravity is useful in finding cavities in soil and rock, depending on the cavity size, its depth below the surface, and the density contrast. An open, air-filled cavity provides the best contrast; a dome cavity in soil overburden that is nearly filled with water or soft clay provides a poor contrast. Gravity is likely to be definitive if $R_a > 1$. Because gravity varies with the distance of the ground surface from the center of the earth, the interpretation of the data is more difficult in hilly and mountainous areas, even when the necessary corrections are made for the differences in ground surface elevation and for latitude.

Gravity exploration is useful in the early stages of site investigation to locate areas that are likely to exhibit significant near-surface dissolution or where there are large shallow ravelling-erosion domes in the soil.

Magnetic surveys similarly sometimes can identify anomalies in limestone formations. However, they generally have not been as useful as gravity.

### 5.6.3 Ground-penetrating Radar

Ground-penetrating radar uses pulses of radio frequency electromagnetic radiation to penetrate materials such as soil and rock and to generate reflections from materials of different density and electromagnetic response (the anomaly). The reflection data include the time elapsed between the emitted pulse and the receipt of the reflection, its return strength, and any phase changes. These data are analyzed and interpreted to find the depth and the lateral extent of the change. The depth of the radar penetration is limited by the power of the impulse and by the attenuation of the electromagnetic energy as it penetrates the ground. In sandy materials above the water table, penetration of more than 100 ft (30 m) is possible; in clays below the water table, only 5 to 20 ft (3 to 6 m) may be possible. Interpretation of the current forms of readout are somewhat subjective and require specialized training and experience. Better methods of processing and displaying the information are being developed that should make this method more useful in the future.

### 5.6.4 Electrical Resistivity

Both electrical resistivity and induced electromagnetic fields measure the electrical conductivity (or its inverse, the resistivity) of the soil and rock. Resistivity utilizes an electric current introduced into the ground by a pair of electrodes, usually at the ground surface. The resulting voltage difference is measured between two or more sensing electrodes that are also placed at the ground surface. Induced conductivity utilizes current flow at depth induced by an electromagnetic field at the ground surface. The effect is sensed at the ground surface from the changes in the strength and phase of the return electrical flux. The electrical conductivity of soils and rocks is related to the ionization of the electrolytes in the soil moisture and to the continuity of that moisture. The conductivity of dry soil or air is low; that of wet soil is high. Rock conductivity depends on the interconnected porosity of the rock as well as electrolytes in the groundwater. Both systems determine the effective conductivity of a significant volume of soil and rock, combining the effects of both low- and high-conductivity zones. The volume tested depends on the spacing and patterns of the electrodes or induction arrays compared to the geometry of the high- and low-conductivity strata or lenses. The depth and geometry of zones of different conductivity are deduced by empirical or mathematical analysis of the data. The nature of the soil and rock within each zone can sometimes be suggested by the computed conductivity (or resistivity), particularly if the geology of the formations underlying the site is known.

Cavities appear in two ways. Above the water table and filled with air, they have no conductivity. Below the water table they are conductive; wet clay filling is usually much more conductive than water filling. There may be little conductivity difference between saturated sand, saturated porous limestone, and a water-filled cavity, however.

As with the other surface geophysical methods, the size of the zone that can be identified depends on depth. The detectable anomaly ratio is usually between 0.5 and 1.5. Unlike seismic methods, described in the next section, shallow strata of high conductivity do not necessarily obscure a deeper stratum of low conductivity. Somewhat better resolution is possible with the electromagnetic induction system, particularly when there are sloping boundaries between different soil and rock formations. The electrical conductivity methods are also useful in estimating the average depth to continuous rock and usually the depth to groundwater. The groundwater is very conductive, while most rock formations, except porous limestones below the water table and saturated mudstones, have low conductivity.

The electrical conductivity is directly related to the potential corrosiveness of the soil and groundwater in contact with steel, such as piping and pile foundations: the higher the conductivity, the greater the corrosion potential of the ground.

### 5.6.5 Seismic Refraction

Seismic refraction usually involves the velocity of travel of a compressive shock wave induced at the ground surface or in a borehole by a hammer blow or a small explosion. The wave velocity depends on the density, modulus of elasticity, and Poisson's ratio of the soil or rock. As long as each successively deeper soil or rock layer has a higher wave transmission velocity, both the wave velocity in each stratum and the depth of the boundary between each can be calculated.

Two wave paths are shown in Fig. 5.6. The shortest is directly along the ground surface from the source of the shock to the detector or *geophone*. A longer path is down to the next stratum, along the surface of that stratum, and back up to the geophone. Although the path is longer, the time it takes for the pulse to reach the geophone could be shorter if the velocity of the wave in the deeper stratum is great and the thickness of the upper stratum is small. By varying the distance between the wave source and the geophone and measuring the time for the first wave to reach the detector, both the wave velocity in the upper stratum and the thickness of that stratum can be calculated.

A modulus of elasticity of each layer can be computed from its wave velocity. However, the modulus varies significantly with the amplitude and frequency of the shock wave. Therefore, the computed elastic modulus only applies to the particular shock wave induced by the test. The modulus that would be of use in foundation design likely will be one or two magnitudes smaller. Therefore, the induced seismic modulus value is of most use in comparing the rigidities of different strata. The type of soil or rock sometimes can be inferred from the shock wave velocity, particularly if the geophysical

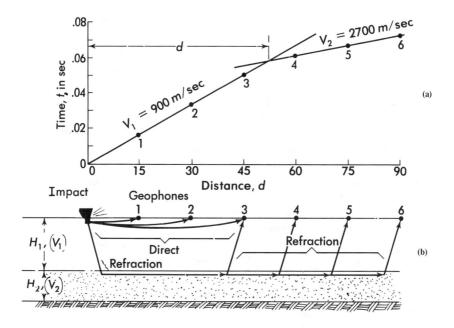

*FIG. 5.6. Simplified Representation of Seismic Refraction, Using Ground-surface Shock Impulses and Ground-surface Geophones or Receptors [after Sowers (1979), p. 320–324]: a. Refraction and Simultaneous Direct and Refracted Signal Paths; b. Velocities of Alternate Wave Paths (Only the Signal that is Recorded First can be Identified; the Later Signals that Arrive at Each Geophone are Obscured by the Continuing Signals from the Wave Train that Arrives First)*

results are correlated empirically with soil and rock test data from a nearby boring.

The method is most effective with strata of uniform thickness, parallel to the ground surface, although mathematical models for interpreting the effects of complex layering with sloping boundaries are available. If the boundaries are irregular, as is the case with a solutioned limestone surface, the data interpretation is highly subjective. When the solution-enlarged slots are wide compared to the depth to the bottom of the slots, the analysis defines a level between the maximum and average slot depth. If the slots are narrow, and the blocks of rock between are wide, the method is likely to identify the average depth to the block tops. Seismic reflection can also find the groundwater level if the seismic velocity below the water table level is greater than that of the unsaturated soil above it. The method seldom can find even large cavities in the overburden because the shock waves through the surface soil travel faster than through the cavity. It cannot find cavities in rock because the higher wave velocity in the sound rock above obscures the slower wave return of the cavity. Advanced dissolution that destroys the continuity of the wave travel in the rock may be reflected in lower rock velocities. In this way, wide pits or large cavities in the rock surface sometimes can be identified. However, if there is even a thin layer of intact rock above the solutioned zone, the lower velocity will be masked by the high velocity above. The ability of the method to sense irregularities decreases with increasing depth below the surface. Typically, an anomaly ratio of less than 1% is necessary.

### 5.6.6 Seismic Reflection

Seismic reflection senses the return of a compression wave from a deeper stratum or body having a different wave transmission velocity. The wave is induced in the ground by a large impact or small explosion at or near the ground surface. The return is sensed by a geophone close to the shock source at the surface, as shown in Fig. 5.7.

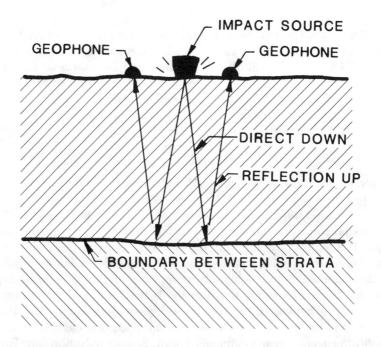

FIG. 5.7.   *Simplified Representation of Seismic Reflection Using Ground-surface Shock Impulses and Ground-surface Geophones or Receptors*

The return is most useful in identifying the depth of very deep boundaries between strata of contrasting seismic velocity. It generally requires an anomaly ratio of 1 or more. The higher the velocity contrast, the better defined the results.

### 5.6.7 Direct Wave Transmission (Uphole, Downhole, and Crosshole)

Uphole, downhole, and crosshole seismic exploration measure the velocity of direct wave transmission between numerous points, both on the ground surface and in boreholes. Both wave source locations and the sensor locations are varied to provide several paths or rays of wave transmission through the body of soil or rock being explored (Fig. 5.8). Typically, several geophones are employed for each wave transmission reception. An air-filled cavity will not transmit the shock wave efficiently; therefore, the wave takes the longer path and the apparent straight line velocity is lower, indicating the cavity. With a number of wave paths, the location of the cavity can be found.

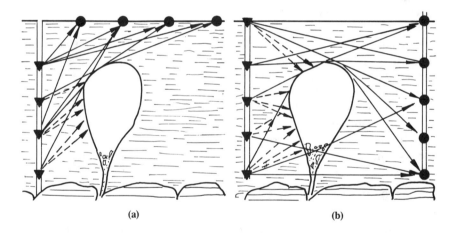

(a)                                                  (b)

*FIG. 5.8. Simplified Representation of Direct Seismic Wave Transmission (the Signals From the Different Wave Paths can be Combined Mathematically, Utilizing the Depths and Directions of the Paths to Determine the Boundaries of the Void that Interrupts the Direct Transmission Paths; the Data are Combined by Tomographic Processing to Develop a Two-dimensional or Three-dimensional View of the Void): a. Uphole Direct Transmission with Sources in the Borehole and Receptors or Geophones on the Ground-surface and Downhole Interchanges Sources and Receptors; b. Crosshole Direct Transmission (the Sources are in One Hole and the Receptors or Geophones in the Other; these can be Reversed to Develop a Clearer Image)*

### 5.6.8 Shear Wave Seismic Exploration

Groundwater, soft to firm clays, and loose, saturated sands have approximately the same compression wave velocity. This makes it difficult or impossible to differentiate between them in procedures utilizing a compression wave as typically produced by a hammer or explosion. However, by utilizing a shear wave impulse and geophones that only detect shear waves, it is often possible to differentiate between them because the soil and rock transmit shear waves but the water does not.

It is difficult to produce large, well-defined shear impulses in the ground and to separate the shear waves from the compression waves in detection. Special equipment is required to generate the shear wave impulse and to detect the ground response. Therefore, this approach is used mostly in very detailed studies.

### 5.6.9 Tomographic Interpretation and Representation

After making large numbers of uphole, downhole, and crosshole measurements, the data can be enhanced using computer-aided tomography (CAT) similar to the CAT Scan X-ray technique of medicine. This makes it possible to construct two- and three-dimensional representations of the anomalies in the ground. The analytical program combines the anomaly signal from each ray path in a three-dimensional matrix using the location and direction of each transmission wave path. It adds the effects of the anomaly as detected from each path and then defines the size and shape of the anomaly in a two- or three-dimensional representation. The multiple-ray exploration and tomographic analysis is usually restricted to critical locations because of the expense of the closely spaced boreholes, multiple shocks, and the volume of data to be processed by the computer.

### 5.7 BOREHOLE PROBE GEOPHYSICS

Borehole probe geophysics was developed in petroleum exploration to sense the response of rock (or soil) in the walls of a borehole. Probe instruments are available for holes as small as 3 to 4 in. (75 to 100 mm) in diameter. Multiple instruments, up to four at a time, are stacked in a single probe that tests the borehole walls in borings up to several miles (kilometers) deep. Most of the instruments measure the qualities of the soil and rock in intervals of 1 to 2 ft (0.3 m to 0.6 m) and usually involve volumes of soil or rock resembling an elongated ball up to approximately 11 in. (280 mm) in diameter. Some instruments, such as the caliper, measure conditions for a distance of a few millimeters. These techniques are utilized to determine detailed variations in soil and rock properties more or less continuously and at great depths below the ground surface. Although low-cost auger or percussion drilling provide the required hole, the equipment mobilization and operating costs can be high. Depending on the access and hole depth, many holes can be logged in 1 day. Therefore, good drill hole planning is required.

The various instruments that have proved useful in geotechnical work and the information gained are given in Table 5.3.

TABLE 5.3.   Downhole Geophysical Logging

| Method | Information |
| --- | --- |
| Caliper: Borehole diameter | Identify very soft soil, unstable closely fractured rock or a cavity intersected by borehole |
| Self Potential: voltage[a] | Correlate strata from one hole to another by voltage pattern comparisons |
| Electrical Resistivity | Electrolytes present, and details of stratification |
| Gamma Ray Emission[b] | Possible clayey seam |
| Gamma Ray Absorption | Density of soil, rock |
| Neutron Absorption | Water content of soil, rock |
| Acoustic Velocity | Modulus of Elasticity at high frequency, low strain |
| Acoustic Televiewer | Image of stratification dip and thickness of thin seams and fissures |
| Temperature | Groundwater circulation |

[a]   Self potential in the ground reflects combinations of electrolytes in the groundwater, as well as cations of the soil and rock solids with different electrical potential. The technique helps identify differences in soil or rock strata and in the dissolved minerals in groundwater.

[b]   Natural gamma ray emission usually reflects the potassium ions in clay minerals. It can help identify thin clay strata or the boundary between a granular overburden and a residual clay in boreholes.

## 5.8 GROUNDWATER EXPLORATION

The previous chapters demonstrated the importance of groundwater piezometric levels and movement in the solution of limestone and the erosion of the soil overburden above. This information is also necessary for evaluating the potential for future development of sinkholes and other water-related site defects, as well as for the planning and design of earth works and structural foundation design.

Observation wells are usually the simplest tool for determining groundwater levels in many locations. The soil and rock boreholes are easily transformed into observation wells by inserting corrosion resistant casings and soil filters, as necessary, below the groundwater level to prevent soil erosion. The casing should be above the ground surface and sealed to the overburden to prevent surface water infiltration. A locked, sturdy cap is needed to prevent damage from the idle curiosity of children and others.

The use of observation wells is complicated in areas of solutioned limestone by the high-conductivity weak zone in the slots or directly on the rock surface. Moreover, the multiple seepage paths in the rock may have several different piezometric levels. Therefore, it may be necessary to have several wells at each location, each sensing a different aquifer. This often requires that an exploratory boring be made at each well location to help define the potential aquifers. Sometimes this is suggested by the water level changes in the hole during drilling. When it is anticipated that boring for soil rock exploration is to be converted to an observation well, either the hole should be drilled dry or any drilling fluid should be the type that decomposes rapidly so as not to restrict seepage into the well.

An observation well is then installed with its well screen filter in each of the suspected aquifers and sealed so that there will be no response to other suspected aquifers. This can be done by drilling separate wells in a cluster with distances between the wells of between 3 to 10 ft (1 to 3 m) depending on the accuracy of the drilling and the fracture patterns in the rock. Alternatively a decked observation well, utilizing a single large hole with several small well casings inside, can be installed. Each smaller well includes a filter screen sealed in a different suspected aquifer so there is no hydraulic connection between them. Because it is often difficult or impossible to seal the decked well properly due to rock fissures and construction difficulties, some specialists prefer separate wells. Electric piezometers, forced into the soil beneath a small borehole, can be utilized to measure multiple water table pressures in soils, especially when continuous readings or remote readings are required. They are more responsive to sudden changes in pressure in soils of low permeability than observation wells.

The groundwater levels are measured at time intervals to determine their response to rainfall infiltration and inflow to and outflow from the site. For rapid changes, continuous recorders may be necessary, particularly for rock piezometric levels, which can change rapidly.

The number and distribution of the observation wells required in limestone formations cannot be determined in advance. Initially, many of the borings should be converted into wells. Based on the aquifers inferred from the boring records and the piezometric levels measured in the first wells installed, additional wells may be necessary to clarify the groundwater patterns.

Contour maps are prepared showing the piezometric levels in each of the different groundwater regimes. This is usually easy for the residual soil overburden. However, for the rock, it can be frustrating when there are several partially independent groundwater systems present. Several alternatives for groundwater contours should be considered, if necessary. Additional observation wells or sealed piezometers may be necessary to resolve inconsistencies.

Trajectories of groundwater flow are drawn perpendicular to the contours. Of course, within the rock, the actual flow paths will follow the rock fissures, typically a zig-zag pattern, but trending in the direction of the trajectories: downgradient.

Groundwater movements can often be verified by introducing a *tracer* such as a dye solution or easily identified, relatively harmless chemical such as salt into an aquifer at a point with a high piezometric level and then attempting to identify the tracer at another point with a lower piezometric level, along a common trajectory. Springs and nearby streams are avenues for groundwater discharge and are often the selected observation points. The tracer is introduced into the aquifer, and observations are made at the discharge point by continuous recording instruments or at intervals based on the estimated time of travel. Measuring at intervals requires some type of absorbing device that accumulates the tracer in between sampling. Dyes such as flurocene are measured by the intensity of their fluorescence under ultraviolet light. Salt concentration can sometimes be sensed by changes in electrical conductivity. Unfortunately, the observed change in the tracer concentration are sometimes much smaller than the similar response to natural materials in the water (i.e., the background response). Tracing groundwater flow in limestones is often fruitless, particularly if the measuring point is another well. The avenues of flow are so convoluted that the test results are inconclusive.

## 5.9 SITE CHARACTERIZATION FOR PLANNING AND DESIGN

Because of the small dimensions of the features peculiar to limestone terrains and the changes in the soil rock profile from location to location, it is usually impossible to develop a single realistic model for the deterministic analysis of bearing capacity and potential settlement, as is often done in foundation engineering. Instead, it is necessary to identify areas of the site with similar conditions. For each area, develop a model, each of which represents the typical conditions, as well as the extremes that are likely to occur in that area. These models focus on the peculiar features of limestone terrains that influence foundation design and performance: the solution depressions, sinkhole, soil ravelling-erosion domes, the inverted soil strength profile, the slot and pinnacle rock surface, and the cavities in the rock. The usual foundation characterizations related to bearing capacity and consolidation settlement are not overlooked. They are considered only after the special concerns for the limestone features are resolved.

## 5.10 DISTRIBUTION OF SOLUTION FEATURES AT A SITE

The number of first-, second-, or third-order streams per unit of land area, compared with the numbers in areas of similar topography nearby, can be found from topographic maps and air photos. The number of solution depressions and sinkholes, and the total areas of solution depressions and sinkholes per unit of land area, are best found from air photos. The larger depressions and sinkholes can also be identified from topographic maps. In addition, some on-the-ground searching may be necessary to identify the smallest sinks and depressions. Maps are then prepared showing areas of similar drainage alignments or drainage density and solution feature densities. These are correlated with the other exploration data to develop the project models for each area. In some sites, the areas of solution similarity are well defined; in others, there may be no pattern of similarities.

In some sites, patterns within patterns can be identified. For example, solution depressions and sinkholes often occur in parallel lines reflecting the joints or strike of dipping formations. The alignment of features along straight or somewhat curved lines are termed *lineaments*. When lineaments cross, it is likely that the intersections will be even more susceptible to solution than elsewhere. This is often reflected in larger or wider depressions at lineament intersections. If there is no depression at an intersection and the conditions in the area are favorable for sinkhole or depression development, the chance of a new one occurring at an intersection are more favorable than elsewhere. In regions of extensive limestone formations, some of the lineaments can extend for miles. Typically, the lineaments are better defined and more closely spaced parallel to the strike in dipping formations (at right angles to the dip azimuth or dip direction). In addition, there will be other directions of more or less parallel lineaments, usually forming rectilinear patterns with occasional random directions in both dipping and level bedded limestone.

The solution pattern approach was used in defining the most suitable area for a large industrial complex in western Kentucky. The area included some of the most closely spaced, well-developed solution depressions and sinkholes in the United States. The number of sinkholes per unit of land area and total sinkhole areas per unit of land area varied greatly over distances of more than 6 mi (10 km). A sufficiently large area with only minor solution development was identified, which coincided with the outcropping of a particular limestone formation. It was selected for the project.

In some areas where limestones are known to exist, solution features are hidden by a covering of deposited soil. In such areas, the site exploration required to determine possible problems is far more extensive and intensive than for areas with visible features such as sinkholes. The work is usually done in phases that combine geophysical exploration and borings in regular patterns such as a simple grid. As information on the nature and extent of the rock defects develops, the data are interpreted tentatively in terms of the effect of such defects on the use of the site. At the end of each phase, the following alternatives are considered: 1) the conditions are so bad that there is little chance of developing the project without extraordinary costs, 2) the conditions appear so favorable that no problems are likely, or 3) some defects have been found but their extent cannot be determined without more data. The third alternative means that decisions should not be made to utilize the site until more detailed investigation has been completed. Several successively more detailed investigations may be necessary before decisions can be made to proceed with the project. Unfortunately, few owners or developers are willing to invest the time and money to develop the required data. Instead, they may assume that sinkhole problems can be easily overcome and will not affect their project costs or schedule. The following example illustrates the danger in this thinking.

A developer of large multi-use projects in Hong Kong leased a tract of low-lying land near the edge of a bay in a previously undeveloped area. The regional geology indicated that the site was underlain by a dolomitic marble whose surface was approximately 60 ft (18 m) below the present ground surface. Between the ground surface was a deposit of soft silty clay alluvium deposited during an era of higher sea level. Limited experience with the foundations of three to four-story buildings in the area showed a few badly solutioned joints. It had been possible to support the structures on drilled shafts by shifting the buildings a little during construction in order to avoid the weaker areas. The new development involved a ground surface rapid transit station and a very large four-story shopping mall covering the entire site. This was topped by four 20-story towers, for offices and for expensive residences.

In order to expedite the completion of the structure, a comprehensive planning-design-construct contract was required by the developer. It included a rigid time for completion, a penalty for delays, and a fixed lump sum price for all work, including site investigation. Before submitting a proposal, each contractor had to make a preliminary site investigation in sufficient detail to estimate what the foundation costs would be. This procedure had been followed in many new Hong Kong projects where most of the sites were underlain by recent sediments over sound granite and granite gneiss. Unfortunately, none of the developers and contractors had experience with foundations on limestone because it occurred only in one localized area and in which there had been no projects of comparable size or cost.

The contractors made a few borings on the proposed dolomitic marble site. Some of the borings found reasonably sound dolomite as thick as 200 ft (60 m) under the 60 ft (18 m) blanket of soft silty clay. However, several foundwater and mud-filled cavities as deep as 200 ft (60 m). Most of the prospective contractors felt they could overcome the problems by shifting foundation locations slightly, and submitted the fixed cost proposals; a few declined to submit proposals because they sensed the possible risks.

An international contractor with extensive experience and financial backing was awarded the contract, starting the time clock. Design was commenced and new borings were made to better define the cavity conditions. As new borings were made, the site looked progressively worse. There were large, deep, interconnected rock cavities in most of the several hundred core borings that eventually were made over a period of approximately 9 months. The cavities were from 165 ft to 330 ft (50 m to 100 m) deep below the ground surface.

A three-dimensional model was developed using 0.25-in. (6-mm) diameter rods set in rigid base. Each rod depicted open and clay-filled cavities, sound rock, and the soft clay overburden. A wide zone of advanced dissolution was identified that covered so much of the site that changes in the structure placement could not avoid significant cavities. Foundations deep enough to generate support below the cavities were beyond the experience of Hong Kong contractors. Equipment and expertise would have to be brought in from other nations. Each foundation would probably terminate at a different level, and measures would have to be devised to cope with the soft clay present in most of the cavities. The owner was adamant that the signed contract could not be modified to include the additional cost and time required. Moreover, the building code was interpreted to require that the foundation design be prepared in advance of construction and not changed significantly during construction without a change in the building permits. The contract was canceled after large financing and development costs had been incurred, and little was recovered by the design-build contractor. This illustrates the fallacy of relying on incomplete data to make plans of unprecedented scale that do not allow flexibility for changes in areas underlain by limestone. This is particularly the case when the rock is hidden by a blanket of more recent deposits that mask any reflection of the limestone defects in the ground surface topography.

The alternative would have involved a study of the potential risks and alternative designs to fit the range in cavity conditions involved. The final construction time and cost could not be definitely known in advance. However, a range in costs and construction time probably could have been developed from the detailed boring data that had been developed in the second phase of the investigation. This would have saved several million dollars in penalties and litigation.

# CHAPTER 6

# SITE PREPARATION AND EARTHWORK

## 6.1 SITE PREPARATION CONSIDERATIONS

Design and construction of foundations in limestone terrain focuses on overcoming the inherent defects and weaknesses of the soil and rock: solution depressions, sinkholes, dome cavities in the soil overburden, slot and pinnacle zones in the soil-rock interface, and collapse-prone caverns in the underlying rock. These structural deficiencies are not static; those involving ravelling and erosion in the soil overburden and the erosion of cavity filling in the rock can change during construction and change greatly within the expected life of most structures. By way of contrast, rock caverns may require centuries to millenniums of exposure to change significantly unless the water is strongly acidic. Existing limestone-related defects can be aggravated by environmental changes generated by construction, such as poor water control, excavation, and imprudent project management. Careless maintenance and waste management once construction has been completed, particularly, bad surface water drainage, and leaking pipes contribute to trouble. Any deficiencies can be aggravated by off-site changes, such as large groundwater level fluctuations or continuous large groundwater drawdown caused by nearby wells. Groundwater pollution can aggravate both soil erosion and rock dissolution. Thus, the problem of foundation design and construction involves much more than the installation of the foundation of the structure; it must involve the total environment that influences the processes that aggravate the limestone and related overburden defects, on-site and adjoining the site on property controlled by others.

Site preparation includes the excavation and filling that usually accompanies construction projects. At sites underlain by limestone, it also includes remedial measures to improve or to correct any existing solution-related defects that might impact the construction and the future structure, as well as measures to minimize the development of new solution defects during the lifetime of the project.

## 6.2 EXCAVATING OVERBURDEN SOIL

Excavation and filling on sites underlain by limestones involve the same engineering considerations required for similar earthwork in other geologic environments, as listed in Table 6.1.

### TABLE 6.1.   Excavation Considerations

| | |
|---|---|
| 1. | Explore known and suspect sinkholes in detail to plan for any required remedial work. |
| 2. | Cut the excavation slopes sufficiently flat to be safe against toppling and sliding. |
| 3. | Remove boulders, pinnacles, and similar obstructions. |
| 4. | Loosen and excavate massive rock. |
| 5. | Control groundwater during excavation by deep sumps or wells that do not induce upward flow towards the excavation surface. |
| 6. | Maintain the stability of the bottom of the excavation against heave and boiling from excessive groundwater pressure. |
| 7. | Evaluate the ability of the soil and rock below the stripping or excavation level to support embankment fill loads. |
| 8. | Determine the suitability of the excavated materials for embankment construction. |
| 9. | Manage the effects of excavation and filling on the overall site environment: a) surface water flow, b) rainfall and surface water infiltration, and c) the groundwater movement, pressure, and chemistry. |

In areas underlain by limestone, many of these considerations are both amplified and somewhat changed by the peculiarities of the residual soil and rock profile. These peculiarities include: 1) soil strength that is higher near the ground surface and which decreases downward toward the limestone; 2) the highly irregular soil-rock interface, particularly the pinnacle-slot irregularities and the soft wet zone; and 3) the multiple groundwater levels, including artesian water pressures from the cavities in the underlying rock.

Sinkholes, solution depressions, and suspect areas often require additional investigation before construction operations commence. The collapse of existing solution domes under the weight of equipment can interrupt construction activities and endanger workers and equipment. Altering surface drainage by excavation can reactivate sinkholes and make repairs difficult. Investigating these suspect areas utilizes the same techniques described in Chapter 5; however, the focus is on low cost and quick and more intensive, localized efforts. The objective is to identify erosion domes, sinkhole throats, and openings in the rock surface that require treatment and to gather information needed to plan remedial work.

Suspect areas can be quickly explored by test pits and trenches to determine if they are underlain by accumulating organic deposits and if there is evidence of a deeper dome or sinkhole throat. Careful excavation in lifts of 1 to 2 ft (300 to 600 mm), accompanied by hand probing the excavation bottom, will disclose loosened soil and organic debris typical of old filled sinkholes and solution depressions. Probing with impact drills can identify deeper weak zones and domes, irregularities in the rock surface, and even deeper openings. These openings can be treated, if necessary, before there is heavy construction or excavation in the area.

Safe excavation slopes are determined using the same methods as in other soil and rock formations: (1) utilize local experience; (2) make stability analyses based on soil strength; or (3) observe the behavior of the slope as it is excavated. However, the stiffness of the residual soil in the upper part of the profile in limestone sometimes misleads both engineers and excavators into adopting steep cut slopes without considering the softer materials that usually lie deeper, immediately above the rock and between the pinnacles. This weak zone can become a surface of local shear and even landsliding into the excavation. Shear at the excavation bottom can propagate upward to encompass the firmer soils above. Because it is difficult to measure the strength of soft zone and to identify its limits, the slope design must be more conservative than in the more usual situation where the entire soil profile can be well-defined before excavation.

Excavating deep into the residual soil or deposited overburden may expose ravelling domes. These are treated in the same way as open sinkholes, as discussed in Section 6.5.

### 6.3 EXCAVATING NEAR THE SOIL-ROCK INTERFACE

Excavation levels that approach the soft, wet horizon or soft soil-filled slot can cause bogging down of rubber-tired and even tracked machines. The heavy equipment can break through firm soil 2 to 4 ft (600 mm to 1200 mm) thick to soft materials below. Rock pinnacles between the slots are obstacles or barriers to excavation equipment movements. Isolated pinnacles, however, are seldom monolithic; they usually resemble stacks of blocks separated by the joints and bedding surface fissures. Once the surrounding residual soil has been removed, such block towers can be readily toppled by a front-end loader or bulldozer. However, this interrupts the normal rhythm of the earthwork, delaying progress and adding to the construction costs.

Excavation within the horizon of alternating pinnacles and slots is time-consuming and expensive. It is difficult to predict the time and the level of work required and the cost. The deepest parts of the pinnacles are wider and more nearly monolithic than the upper part, as shown in Fig. 2.6, and

sometimes cannot be broken by excavating machines. High-impact demo-
lition points and blasting may be required to fragment them for removal.
Although the author has seen no data that show that the rock mass has been
significantly damaged by blasting pinnacles, it is prudent to use multiple
detonation delays so as to limit the ground response to a peak particle
velocity of 2 in. per sec. This is the same limit imposed to prevent damage
to plaster in buildings. By way of contrast, the soft, wet materials in the slots
sometimes must be bailed out using buckets designed for muck. The pinnacle-slot zone is neither soil nor rock; instead, it is the worst combination
of both and is usually much more time-consuming and expensive to excavate than either alone.

## 6.4 EFFECT OF SITE EXCAVATION ON SINKHOLE ACTIVITY

Site excavation has three effects on the soil remaining above the limestone: (1) The vertical stress in the remaining soil due to its own weight is
reduced; (2) the vertical stress in the remaining soil due to the weight of new
structures at the ground surface is increased; and (3) the seepage path from
the ground surface to the soft zone and rock is reduced. All of these have an
impact on future sinkhole development.

Reducing the vertical stress can be detrimental when piezometric pressure in the rock voids connecting with the soil rock interface exceeds the
weight of the soil overburden. Local heave of the soil overburden can
develop, possibly accompanied by mud boils. If the overburden is very
sandy, sand boils and quicksand can develop. As a further complication, the
onset of boiling may be temporarily delayed because of the restraining effect
of closely spaced pinnacles. This may encourage deeper excavation before
water control measures are undertaken. The delay in water control is often
followed by spectacular fountains and boils that will halt construction and
seriously damage the site. If the water pressure in the rock drops sufficiently
from excavation drainage, downward seepage erosion is aggravated, accompanied by the development of erosion domes.

When fountains and boils appear, it is necessary to stop further excavation until the water pressure can be relieved by pumping from wells that
extend into the rock cavities. The consequences of renewed ravelling erosion and dome development and of site settlement from relieving the water
pressure must be weighed against the need for deeper excavation.

When the site exploration data and the anticipated weather during
construction indicate that the water pressure in the rock could become high,
it is prudent to install piezometers in the rock cavities before proceeding with
construction so as to verify suspected pressure increases and to monitor the
pressure changes caused by weather, as well as those produced by construction operations. With good current piezometer data, steps can be taken to

prevent heave and boiling before it occurs, making it unnecessary to use costly emergency procedures in the middle of excavation work.

The shorter seepage path to rock voids caused by site excavation can increase seepage erosion in the soil because the densest, most impervious soils are in the upper part of the profile. Excavation of the stiff overburden above erosion domes that have propagated up to the stiffest upper strata can trigger local collapse.

Excavation reduces the thickness of the hardest soils in the profile and increases the vertical stresses in the deeper softer soils caused by surface loads imposed by future structures. Therefore, more foundation settlement sometimes develops despite the unloading effect of the excavation; any excavation planned should consider the increased sinkhole activity and increased settlement compared to the benefits.

Excavation reduces the soil cover above small sinkholes and makes them more vulnerable to collapse because of the shorter seepage path and loss of the stiffer, protecting near-surface strata. Proof-rolling with a fully loaded 50- to 100-ton roller can detect domes near the excavated face. During rolling, the ground is watched for locally increasing deflection from the slowly moving loaded roller. Ordinarily three to four passes of the roller are a reasonable test. When the heavy roller is not available, a fully loaded heavy earth mover can be used; however its load concentration on each tire is not as great as that of the roller.

Deep excavation at the site makes it possible to sense smaller erosion domes near the rock surface. Therefore, if high structural loads are to be imposed on the remaining overburden, additional geophysical exploration after excavation will be prudent.

## 6.5 OPEN SINKHOLE TREATMENT

Open sinkholes, including domes exposed during site excavation, obviously require treatment if they cannot be avoided by shifting the structure away from them. The most effective approach is to excavate into the narrowest point of the bottom, the *throat* or opening, to expose the limestone pinnacles and the rock surrounding the throat. A *plug* (i.e., a structural blockage) is then constructed to bridge across the opening and transfer the load of future backfill to the rock (Fig. 6.1). Clay adhering to the rock is removed insofar as possible, so as to permit good contact and some bonding between the plug and the rock (Fig. 6.1a). The most secure plug is concrete, approximately 1.5 times deeper than the width of the throat (Fig. 6.1b). If the throat is very wide, a thinner, reinforced concrete plug can be constructed to save concrete, but at the cost of more labor and placing the reinforcement in the constricted space. Both the plain or reinforced concrete plugs are very effective in preventing further erosion at that location and in providing structural support for the overburden loads.

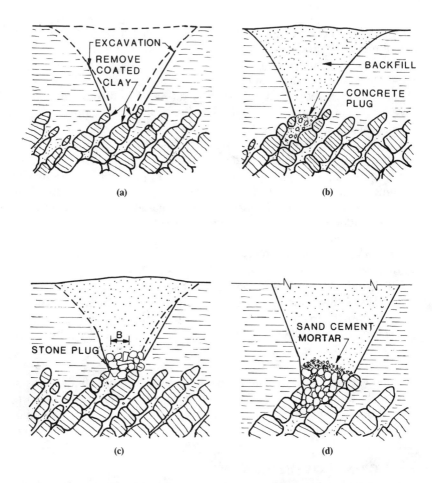

FIG. 6.1. *Sinkhole Throat Filling: a. Cleaning the Narrowing Rock Throat of a Sinkhole and Removing the Clay Coating on the Rock; b. Non-reinforced Concrete Plug, Height, H = 1.5 Times the Width, B of the Narrowest Point of the Throat; c. Rock Fill Plug, with the Diameter of the Deeper Rock Pieces Greater than Approximately One-half of the Throat Width, B; d. Partially Grouted Rock Fill Using Rock Smaller than One-half the Throat Width, B*

When there is considerable downward infiltration, blocking downward seepage at one point can aggravate ravelling and erosion and new sinkhole activity nearby. A rock fill plug (Fig. 6.1c and d) is an acceptable alternate. Rock blocks or boulders wider than about half the rock opening width dumped into the hole will arch across the bottom opening. These are followed by successively smaller rock particles (Fig. 6.1c), forming a graded filter. Although rounded boulders and gravel can be used, angular quarry rock tends to wedge in place more effectively. If sufficiently large rock is not available, the largest rock available is dumped into the hole until its surface is above the narrow throat at a height of approximately two-thirds the width of the opening. Sand cement mortar is then placed on top of the rock. The mortar should be fluid enough that it will penetrate the larger voids and bond the rock pieces fragments together for a depth equal to the rock fill height above the narrowest point of the natural opening. In this way, a crude, grouted-stone, domed plug is created within the sinkhole throat (Fig. 6.1d). It is not necessary to completely fill the voids unless a relative impervious plug is required.

A pervious plug (Fig. 6.1c) permits downward seepage and drains water from the overburden if considerable surface infiltration is anticipated. The grouted stone can be made to be pervious by reducing the grout slump and volume somewhat. It should be tested after construction. Alternatively, short pipes leading to a source of water can be inserted through the fill rock before grouting. A graded filter is required to resist the downward erosion of soil. One alternative is a graded stone-to-sand filter that meets the usual criteria for filtering soil fines (Sowers, 1979, pp. 119–20). An acceptable alternative is a heavyweight geotextile filter.

If it is not practical to excavate into the throat and clean the rock, *compaction grouting* can often form a satisfactory relatively impervious plug, depending on the thickness and stiffness of any clay coating on the rock in the throat. Compaction grouting utilizes pumped cement, sand, and, sometimes, pea gravel mortar with a slump of 3 to 5 in. (75 to 125 mm) at pressures from 250 to 500 psi (1.7 to 3.4 MPa). The high-pressure mortar or grout will displace soft soil in narrow openings and will develop a stable and often watertight plug (Fig. 6.2). If the rock is coated with clay, the plug may not remain water-tight, and may become loosened, if the clay eventually erodes. Because it is rarely possible to measure the compaction grout plug dimensions (except by estimation from the volume of grout), pumping a generous volume of grout is appropriate until the hypothetical plug height is 2 to 3 times the opening width. Once plugged, the remainder of the sinkhole can be filled with materials compatible with the future load on the fill surface. Because of the limited space in the deeper portion of the sinkhole, compaction of cohesive soil filling may not be practical or economical. Crushed stone, dumped in place, will settle slightly under static load. This may be tolerable for some situations. However, the settlement can be significant under vibration or earthquakes. Crushed stone, gravel, and sand filling can be vibrated and densified satisfactorily even in narrow

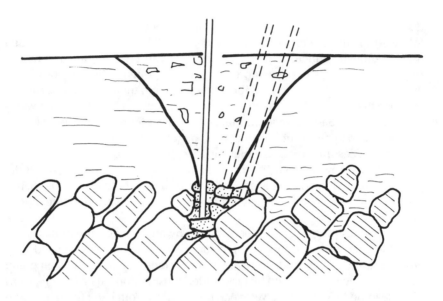

*FIG. 6.2. Compaction Grout Plug in a Deep Narrow Sinkhole Throat*

openings. Fill concrete or pourable fill (fluid low-strength concrete) are stable under both static and vibrating loading, but are more expensive.

When the working space for sinkhole filling is restricted, such as between buildings or under an existing structure, it is often expedient to utilize compaction grouting for the plug, followed by low-pressure grouting of the remainder of the opening, so as to minimize ground settlement. The fill grout includes the greatest sand content that is compatible with pumping in order to limit its shrinkage.

The grout pressures for filling an erosion dome (or the soil debris that cannot be removed) is controlled to avoid heave of the ground surface from the pressure. A typical safe pressure is approximately 1 lb per sq in. for each foot (23 KPa for each meter) of depth below the ground surface. Larger pressures may be possible, but must be accompanied by careful level measurements of the ground surface. Grouting is stopped at the first sign of heave, or when the grout pressure rises above the limiting pressure during continuous pumping.

When the infilling of a sinkhole throat is too stiff to displace with high pressure, a more effective, but expensive technique, *jet grouting,* may be successful, as described by Kauschinger and Welsh (1989). This process involves pumping a fluid grout into the soil with a rotating high pressure jet. The jet erodes soil and cuts stiff clays and soft erodible rock into gravel to small boulder-sized pieces. Pressures of 4,000 to 7,000 psi (30 to 50 MPa) are typical at the grout nozzle; however, that pressure dissipates rapidly within the soil and does not cause heave when the volume of the grout is properly controlled. The larger particles of soil, including sand and gravel in the sinkhole filling, mix with the grout, producing a mixed-in-place concrete. The largest particles, including boulders and stiff clay chunks, are encapsulated by the mix. If the sinkhole infilling consists mostly of stiff clay with gravel and rock fragments, the clay can be first cut into small pieces and partially eroded into a slurry by high-pressure water or air-water jets that are part of the more advanced jet equipment. In this case, the clay chunks and slurry are flushed upward out of the ground. This is immediately followed by or alternated with cement mortar jetting, which binds the remaining larger, rock-like particles and some larger clay chunks to plug the opening. The success of this method depends on the versatility of the equipment and the skill and experience of the operators. Some trial and adjustment will be necessary before the process is successful in each different application. When clayey fines are washed to the surface in the form of a slurry, their disposal presents an environmental disposal problem and a significant cost.

### 6.6 SOLUTION DEPRESSION TREATMENT

Shallow solution depressions in deep stiff overburden above areas of rock dissolution but without erosion cavities in the overburden usually can support light one- and two-story structures, such as houses and store buildings. However, it is imperative to replace any soft or compressible surficial accumulations in solution depressions with compacted fill. Alternatively, preloading can stabilize the weaker or more compressible materials (Fig. 6.3). This should be done only if sufficient time for soil consolidation is available and if a future structure supported on that fill can tolerate some small, continuing settlement. However, if the infilling is highly organic, there will be some long-term settlement despite preconsolidation. Measures should be taken to prevent water infiltrating through the soil and initiating erosion dome development. This surficial treatment is prudent where there is little history of dome erosion and sinkhole development in the area.

When the compressible infilling is deep, some constructors have recommended wick drains of plastic and geotextile to accelerate consolidation. However, such drains could accelerate deep infiltration from surface runoff and thereby accelerate any potential sinkhole activity. This is not recommended.

### 6.7 IMPROVING OVERBURDEN RESISTANCE TO DOME COLLAPSE

Despite the most intensive exploration, it is possible that unidentified erosion domes may be present and that new ones could develop. While

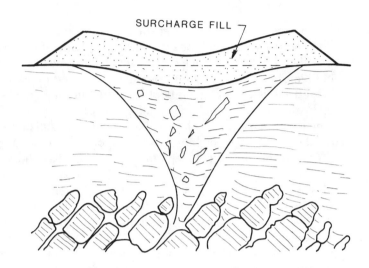

*FIG. 6.3.   Preloading Shallow Organic Debris and Soft Clay in a Filled Sinkhole or Solution Depression*

experience demonstrates that the risk is very small, there are additional measures in site preparation that may reduce the risk further, although it may be difficult to justify them economically.

If a dome is suspected or positively identified, it can be treated in the same way as a sinkhole. Alternatively, a hole can be drilled through its roof and the cavity filled with high slump fill concrete or similar materials (Fig. 6.4). Filling without a positive seal in the throat at the soil-rock interface will not always prevent future enlargement of the cavity, although the rate of erosion will be greatly reduced. However, in most cases, it will be stopped unless there is some severe environmental change in the future.

A second approach has been to *precollapse* the erosion domes during site preparation. Three methods have been used. The oldest is to utilize explosives. This has sometimes been successful, if the residual soil or deposited soil overburden has little or no cohesion. The procedure requires experience. Typically, holes are drilled on a grid pattern with spacings from as little as 10 ft (3 m) and as great as 30 ft (9 m). Explosives are placed in each hole, often alternating an explosive charge with decking of an inert material, such as sand. Too much explosive can lift and loosen the soil mass; too little will not be effective. The domes collapse and become filled with loose soil. The surrounding intact soils may be somewhat densified by the concussion; however, they are sometimes loosened if the explosive charge is too large. Surface water infiltration may be increased and a new dome may eventually form at the point of detonation because the soil has been dis-

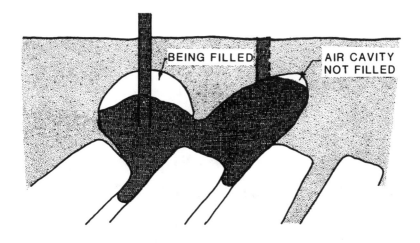

FIG. 6.4. Dome Filling with Flowable Fill (Weak, High Slump, Fine Aggregate Mortar) or High Slump Fill Concrete

rupted and loosened by the explosion. This method should be considered a
temporary measure to be used only in a sandy overburden. In some cases,
it has eventually increased the frequency of sinkhole development.

A second method is to collapse the dome with high-impact compaction
by repeatedly dropping a weight of tens of tons from heights of up to 75 ft
(22 m) (Fig. 6.5). The copyrighted term for this is *Dynamic Compaction*. A
more descriptive term is *high-impact compaction*. The process not only
collapses the dome but also densifies the collapsed soil and the surrounding
and deeper soil to varying degrees. However, it does not correct the underlying soil erosion and ravelling problem; therefore, the dome formation
process can begin again if the environmental conditions are favorable. The
reactivation of ravelling and erosion may be retarded because impact on the
undisturbed soil increases the soil density and reduces the hydraulic conductivity of soils that were naturally loose. Before this process is utilized, the
effects of soil compaction on the undisturbed soil should be evaluated by
full-scale field testing. The method appears to be most effective in sandy soils
as a means of retarding but not necessarily preventing sinkhole development.

A third and technically sounder method is to use *stone columns:* crushed
stone cylinders installed by impact and vibratory action (Fig. 6.6). These
columns, 2 to 4 ft (600 mm to 1200 mm) in diameter and 6 to 12 ft (2 to 4 m)
apart, densify sandy soils, as well as collapse any domes they intersect.
Although they will collapse a dome in clayey soils, they are not very effective

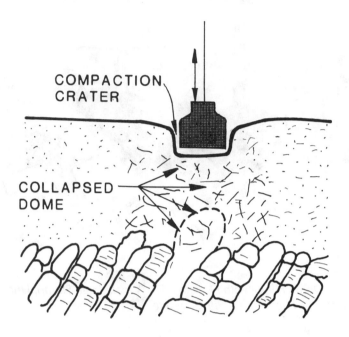

FIG. 6.5.   Intentional Preconstruction Collapse of an Erosion Dome in Soil
Overburden by High-impact Compaction

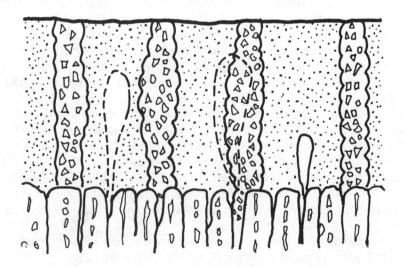

FIG. 6.6.   Stone Columns Inducing Dome Collapse, Providing Drainage,
and Increasing the Soil Density in a Sandy Overburden

in compacting the stiff clayey residual soil blocks that collapse into the dome. The stone columns can help to minimize future erosion by forming a filtered vertical drain column that concentrates seepage drainage into the erosion-resistant column. This requires utilizing an aggregate in the stone column filling that effectively filters the surrounding soil. Stone columns have proved successful where the limestone has significant primary porosity, low secondary porosity, and the overburden is sandy. Where there are localized openings, such as sinkhole throats in the rock surface that can be identified from the ground surface, a column should aim for the dome throat so that any drainage is focused on an existing rock cavity. If the opening in the rock surface is large, the vibrator and the column may drop downward into a cavity below. While this may be beneficial in inhibiting further sinkhole development at that specific location, a large amount of crushed stone may be consumed in filling the rock cavity before the dome is filled. Moreover, filling a cavity in the rock below may partially restrict groundwater flow in the cavity and initiate sinkhole activity elsewhere. The stone columns appear to be most effective when the overburden is sandy.

## 6.8 INHIBITING RAVELLING AND EROSION AT THE SOIL-ROCK INTERFACE

If the soft zone is widespread with numerous small erosion domes above, and there are few or no pinnacles, further dome erosion can be minimized by injecting viscous grout or high slump sand-fly ash-cement mortar to form a more-or-less continuous barrier on top of the rock. This is termed *cap grouting* (Fig. 6.7).

The grout holes are laid out in a grid, with spacings from 10 to 20 ft (3 to 6 m), depending on the estimated space between the larger voids in the rock surface and any soft zones in the soil-rock interface. The grout pressure measured at the ground surface is equal to 1 to 1.5 times that of the overburden stress at the soft zone level. It should be enough to force the soft soil into open rock voids below, replacing the soft soil with the grout. During grout injection, the ground surface elevation is observed for heave. If this occurs, injection is stopped immediately in that location. It is resumed in the next adjacent ungrouted hole. If ground surface heave occurs, the pressure is reduced until the grout goes in without heave. After the holes in the grid appear to be filled, new holes are drilled midway between the initial or primary holes and attempts are made to inject grout. If little or no grout is injected in these secondary holes, the cap is considered to be complete in that area. When significant volumes of grout are injected in the secondary holes (comparable to that in the primary holes), additional intermediate holes are grouted. When a large number of secondary holes in an area take significant amounts of grout, tertiary holes may be needed between the

## CLOSELY SPACED GROUT HOLES

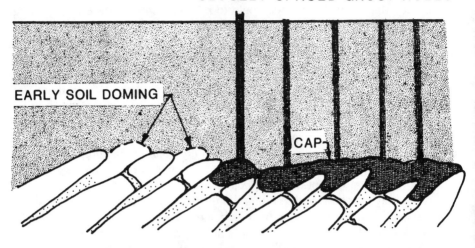

*FIG. 6.7. Cap Grouting of the Tops of Narrow Slots and Small Cavities Using Stiff Mortar to Prevent Soil Erosion at the Soil-rock Interface*

secondary holes that took grout. Also in any areas that exhibited significant heave, intermediate holes are grouted to fill any new cavities caused by the heave.

Sometimes the stiffer compaction grouting procedure is employed for cap grouting. It is effective in blocking large voids in the rock surface when these can be pinpointed; however, the stiff, very viscous grout usually does not travel far enough through the soft soil to fill small voids between widely spaced boreholes and to block smaller holes in the rock surface. If cap grouting is done using compaction grout pressures, heave can occur at the ground surface with the danger of cracking the soil overburden and defeating the cap-grouting objectives.

Although grout strength is not important (750 psi or 5 MPa is adequate), the grout or mortar should set rapidly so that it will not flow indefinitely into rock cavities. If large volumes are pumped in without building up pressure, grouting is stopped after a predetermined volume, such as 100 ft$^3$ or m$^3$, has been pumped. The grout is allowed to set, hopefully blocking some of the large open voids in the rock-soil interface that are accepting the grout. Grouting is then resumed until the pressure builds up to the levels described previously. If the soft zone is extensive, several split spacings of grout holes may be necessary to seal the rock surface.

The procedure is controversial. There are limited data that suggest that sites that have been cap grouted are less prone to erosion dome enlargement and collapse than similar nearby sites. However, it is also plausible that the structure built above the cap grouting is responsible for the reduced dome erosion development by reducing surface water infiltration. Where the ground surface is protected against water infiltration, cap grouting may be only an added, but indeterminate, enhancement of safety at a significant price.

If there is no well-defined soft zone, it is sometimes possible to grout some of the small cavities in the upper few feet of the underlying rock to prevent downward erosion (Fig. 6.8). This is difficult, expensive, and often impossible because the locations of these cavities cannot be determined accurately from the ground surface. Even closely spaced grout holes may not find a particular cavity that might be responsible for future erosion. Sometimes attempts are made to fill all the near-surface cavities in the underlying rock to inhibit further erosion and ravelling development. While this may be theoretically helpful, it may be both impossible and potentially detrimental. Blocking flow in cavities can increase the piezometric levels in the rock. This may cause the water pressures upstream to increase during wet weather and force water into the overburden above. This can enhance ravelling and erosion when the water pressure subsides during dry weather. If there is

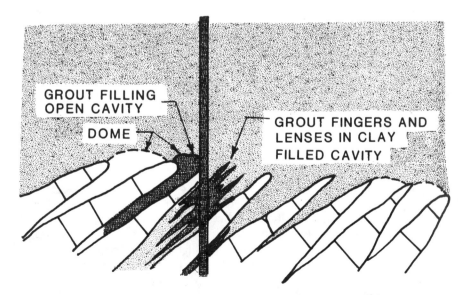

FIG. 6.8.   Selective Grouting of Rock Cavities and Slots Close to the Rock Surface

ow in the cavities, the grout is washed away as it is
fort is worthless.

s observed significant amounts of diluted grout in springs
r as 1 mi (1.6 km) from grout injections in limestone
cavities. In such cases, resuming grouting with a much more viscous and
quick-setting grout is usually more successful in blocking flow. At best,
trying to fill rock cavities when water is flowing is a difficult trial-and-learn
process that is peculiar to each specific site. Experience elsewhere can be
misleading. Instead, an intimate knowledge of the rock cavities, plus a vivid
imagination on the part of those doing and supervising the grouting, is
essential.

## 6.9 EMBANKMENT CONSTRUCTION

Embankment construction is done in the same manner as for other site
development projects. The original ground surface is prepared by *stripping*
(removing) topsoil and *grubbing* to remove roots, boulders, and other surface
debris to the extent required for the ultimate purpose of the fill. Placing a
relatively impervious embankment on residual or deposited soil over limestone foundation inhibits future dome formation and rock solution at the soil
rock interface by reducing surface water infiltration. However, to be most
effective, the fill surface should be shaped to shed water rapidly, both during
and after construction, and to dispose of the surface runoff water so that it
will not create new erosion dome action elsewhere. Special attention is paid
to surface drainage and infiltration. Altering the surface topography can
redirect water flow, concentrating water and aggravating downward seepage
and consequent ravelling, erosion, and dome development elsewhere including on property owned by others. Unfortunately, attention is seldom
directed to this problem until damage appears.

If the fill soil is more pervious than the natural ground beneath, future
infiltration may not be inhibited and ravelling dome development can
develop beneath the embankment. When other environmental conditions
are conducive to ravelling dome development, the fill can aggravate or even
cause erosion dome development. For example, waste gypsum impound-
ment-fills, known as *gypstacks*, have been examples of focused, aggravated
dissolution. Embankments higher than 100 ft (30 m) are constructed of waste
calcium sulfate (gypsum) that has been settled out of suspension after
treating phosphate ore with sulfuric acid. Initially, the stack is constructed by
building dikes approximately 15 ft (5 m) high, thereby creating a settling
pond. The dilute sulfuric acid is recovered as the suspended gypsum settles
out and is recycled, leaving the waste in the form of silt to sand-sized
particles with some sulfuric acid in the voids. This eventually partially
cements into a weak, porous rock-like mass. The precipitated calcium

sulfate or gypsum is excavated and used as fill to raise the dikes and reestablish the settling pond. However, the dilute acid in the voids sometimes leaches into the soil overburden. Eventually, the infiltrating acid may reach the limestone beneath and cause aggravated dissolution. In at least two cases known to the author, sinkholes wider than 75 ft (23 m) developed in the gypstack surfaces, as shown in Fig. 6.9. In a similar failure, the acid dissolved a solution pit in the underlying porous limestone more than 100 ft (30 m) deep and more than 100 ft (30 m) in diameter. This was corrected by compaction-grouting the pit bottom and then filling the remainder of the hole with clayey soil. Such infiltration of water or waste through an embankment can be prevented by providing an impervious membrane of compacted clay or plastic beneath the embankment. The membrane should be mounded at the center to drain any infiltration to the perimeter where it is collected and disposed of in a way that it presents no hazard to the embankment or the surrounding environment.

### 6.10 RESIDUAL SOILS FOR EMBANKMENT CONSTRUCTION

The excavated residual soils are evaluated for use in embankment construction by the conventional laboratory tests for soil classification, by

*FIG. 6.9.  A Sinkhole 75-ft (23-m) Wide in the Top of a Gypstack 125 ft (38 m) high in Central Florida*

their moisture-density relationship upon compaction (Sowers, 1979, Chapter 6), and, when appropriate, by strength and consolidation tests. The gravelly sandy clay residual soils derived from many limestones usually make good embankments for supporting pavements and structures or for the water-retaining portions of earth dams and lagoons. However, there are three conditions that sometimes prove troublesome. First, the residual clays occasionally exhibit high plasticity and, therefore, are prone to develop swelling pressures. The compacted highly plastic clays heave upon wetting and developing shrinkage fissures on drying. Second, local coarse chert concentrations and occasional limestone boulders derived from the pinnacles may have to be removed in order to compact the soils adequately. Third, soils from the soft wet zone are difficult to dry out and usually must be wasted.

### 6.11 AN EXAMPLE OF MISDIRECTED SITE PREPARATION AND CONSTRUCTION

Several decades ago, a small city undertook a major addition to its water treatment plant. A new plant was planned adjacent to the old one, on a low hill of residual clay over dipping limestone. The site preparation required excavating up to 20 ft (6 m) of residual soil to level the site. An additional 20 ft (6 m) more was excavated for construction of new reinforced concrete sedimentation basins and a clear well. A reputable designer was retained to prepare the project design, but not to provide on-site monitoring or inspection of the construction.

In order to save money, the city provided the site investigation work as well as the construction inspection. This was accepted by the engineer although the city did not specify the qualifications of those who were to provide those services. The city borrowed a drilling rig from the highway department and hired a former agricultural soils expert to supervise the site exploration. The borings were made with a truck-mounted power auger. The auger cuttings were examined by the agricultural expert and classified in terms of its texture, origin, and pedologic profile (the upper 3 to 5 ft (1 to 1.5 m)). Most of the borings extended to 30-ft depths where further progress was impossible with the auger. This was interpreted as being the bedrock because limestone fragments were recovered in the deepest auger cuttings. Although no intact or undisturbed soil samples were made and no engineering tests were made, the expert reported stiff soil 30-ft deep over limestone. This was apparently based on the driller's record of drilling progress. On this basis, the engineer designed a continuous mat foundation for the deep basins supported on a levelling layer of crushed stone on top of the limestone.

Construction commenced with general excavation up to 10 ft. During this work, a bulldozer broke a 12-in. (300-mm) water pipe leading to a 50,000-gal (190,000-l) overhead tank. The entire tank emptied and the water disappeared into the ground. The city inspector was delighted that the water had disappeared because the excavation work could continue without muddy conditions.

A second suspicious event occurred while excavating 20 ft (6 m) deeper for the sedimentation basin. Shallow rock knobs were encountered with soft soil between. The knobs were blasted off and the soft spots were filled in with crushed stone. The inspector was pleased because the amount of blasting was limited and the gravel made it possible for the excavating equipment to operate, except at one point where a bulldozer sank several feet into soft clay below the level of the mat foundation. A crane was required to lift it out. However, more crushed stone made that spot stable.

The plant was completed with no further problems. However, during start-up, some minor problems developed. The doors of the operating building adjacent to the basins began to stick making it necessary to cut off their tops at an angle so they could be closed. Diagonal fissures appeared in the junction between the sedimentation basin and the clear well. The design engineer was consulted. He suggested that the fissures were caused by the non-uniform support of the basins: partially on crushed rock on top of limestone and partially on crushed rock on clay. He believed that the structure could absorb the differential movements without detriment to the plant function and suggested that the plant be put into operation.

The basins were filled and the plant started. The following day, the plant operator heard a loud "gurgling noise" as he described it. Immediately afterward, the operating building shuddered and tilted downward toward the adjoining sedimentation basin. The operator stepped outside and saw that the basin had settled and the water in it was draining out. When all the water had disappeared, holes were seen in the bottom (Fig. 6.10a). The outside of the operating building, which had tilted toward the basin, exhibited a large diagonal fissure and the distorted door frame (Fig. 6.10b).

The subsequent investigation disclosed a deeply slotted bedrock surface with both stiff clay and soft clay filling. Apparently, the water tank draining on the partially excavated site had developed (or aggravated) several erosion domes between limestone pinnacles. Small differential settlements of the basins may have caused cracking, which caused leaks once the basins were filled with water. The final basin failure was attributed to erosion dome collapse in the residual clays between several of the larger slots. Eventually, the problem was corrected by extensive, expensive grouting. The corrective work was far more costly than the money saved by the city in its exploration and inspection work.

(a)

(b)

*FIG. 6.10. Damage to Water Treatment Plant Structures from Sinkhole Activity Created by Excessive Site Excavation, a Water Main Fracture, and Construction Mismanagement. a. Bottom of the Sedimentation Basin with Dropouts. b. Cracking in the Operating Building with the Sedimentation Basins on the Left*

### 6.12 MINIMIZING EARTHWORK DISPUTES

Because of the inverted residual soil strength profile, the multiple and often changing water tables, and the irregular but abrupt soil-rock interface, it is sometimes necessary to make changes in the project layout and in the design of the structures. Different construction equipment and procedures may be required that had not been anticipated. This causes lost time and additional expenses for the project owner-sponsor, the designers, and the constructors. Too often, much more time and money is lost in determining who is to blame (and required to pay) than in correcting the technical problem quickly and efficiently. The procedures suggested in Table 6.2 can minimize project delays and costs.

TABLE 6.2. Procedures for Avoiding Delay and Conflicts in Resolving Unforeseen Problems in Earthwork Construction

| | |
|---|---|
| 1. | The project owner or sponsor will acknowledge that the very nature of a site underlain by limestone involves the risk of unforeseen (and sometimes unforeseeable) problems. Because it is the owner's property, the owner has primary responsibility for the risk of unforeseen site conditions. |
| 2. | In order to minimize the unforeseen, the owner or sponsor will expend a reasonable effort in site geotechnical investigation. The reasonable effort for a limestone site will significantly exceed that expended for similar projects in areas not underlain by limestone. |
| 3. | All investigation information, including the geological and geotechnical interpretations, will be provided to the designers and constructors. What constitutes fact and what constitutes interpretation of the facts should be clearly differentiated in the investigation documents and in the project specifications. However, the designers and constructors will make their own interpretations if they see fit to do so based on their experience and technical training. |
| 4. | The same professionals involved in project planning and site exploration will be involved in monitoring or inspecting site preparation, excavation, drainage, and foundation construction in order to benefit from the continuing accumulation of knowledge approximately the ground conditions at the site. |
| 5. | Unforeseen conditions will be revealed to all parties as soon as they are encountered. If possible, all parties should agree on the remedial measures to be taken. However, immediate action is more important than agreement because problems usually become worse with the passage of time. Steps will be taken to remedy the problem promptly and efficiently. Cost and time records of the remedial work will be kept. |
| 6. | A deviation/dispute mediator will be retained by the owner before the construction contract is signed. This person must have a background that includes engineering, geology, and law, as well as experience in construction in limestone terrains. This person must be acceptable to all the parties involved in the work. The mediator will verify that all parties are currently informed approximately the work progress and particularly about problems that arise in excavation, filling drainage, and foundation work. The mediator will promptly mediate disputes that arise in determining the methods correction and of proceeding with construction and how any additional time and costs should be allocated. |

It is possible that such a mediator will make mistakes. However, the author has observed the cost of the mediator's mistakes is far less than the cost of prolonged arguments and unending litigation.

# CHAPTER 7

# FOUNDATIONS SUPPORTED ON OVERBURDEN

## 7.1 FOUNDATION ALTERNATIVES FOR SOIL OVERBURDEN

The design of foundations in limestone terrain is controlled by the same considerations of minimum depth below the ground surface, bearing capacity, and potential settlement as for foundations in other materials, (Sowers 1979, pp. 443–444). However, these considerations are prefaced by additional criteria related to the risk of subsurface *subsidence* (i.e., slowly developing local settlement), catastrophic ravelling-erosion dome collapse, and, rarely, collapse of rock cavities below. These limestone considerations are considered first; the usual foundation design criteria apply after these special problems are resolved.

The major foundation design alternatives are listed in Table 7.1.

TABLE 7.1.   Alternative Foundation Designs

| | |
|---|---|
| 1. | Shallow foundations supported on the residual (or deposited) soil overburden or on fill supported by the soil overburden. Improving the overburden may be desirable or necessary, as discussed in Chapter 6. |
| 2. | Foundations supported directly on the limestone where the overburden is thin and the rock is either competent or its defects have been remedied. |
| 3. | Deep foundations extending through the overburden and solutioned rock to support on more competent rock. |

Some projects require a combination of these alternatives, depending on the variability of the ground and the tolerance of different components of the structural systems to ground movement. In some cases, it is necessary to change the location and configuration of the structures so that they adapt to the differences in the soil and rock integrity from point to point and area to area. For example, a high-rise structure with a small footprint and concentrated high foundation loads might fit a small area of the site that is cavity-free. A single-story structure with the same floor space might cover such a large area that it might extend into portions of the site with severe solution-related defects. Although the vertical stress increase in the foundation soil caused by the one-story building would be only a fraction of that of the multi-story building, the larger area encompassing solution-related defects would be of far more concern despite its lower foundation stresses. More-over, environmental changes, such as leaking pipes or downspouts concen-trating water at the surface, could make a one-story structure, partially supported on sinkhole-prone ground, a far greater risk than the heavier multi-story building on a small area having no solution-related defects. In either case, environmental controls are necessary to minimize future lime-stone deterioration.

A structure could be changed so it adapts to poor soil and rock condi-tions: making it resistant to potential dome collapse. Even if dome collapse is only a remote possibility it might be prudent to design the structure to be tolerant to the usually small differential settlements that occur from overbur-den sag and pinnacle punching.

## 7.2 BASIS FOR DESIGN OF FOUNDATIONS SUPPORTED BY RESIDUAL SOIL

The analysis of foundations on the residual soil or on deposited soil (or fill) overlying the residual soil begins with an evaluation of the effect of the inverted soil strength profile. The stiffest, most incompressible soils are near the ground surface. With increasing depth, there may be small variations in the residual soil strength and stiffness, but no major changes until close to the rock surface. The rock surface is irregular with rigid rock pinnacles penetrating into the firm to stiff residual soil. The impaled soils may become even stiffer where they are confined laterally between the pinnacles. Be-neath the firm to stiff soils impaled between the pinnacles, there may be very soft, underconsolidated clays. When there are no pinnacles, or the pinnacles are widely spaced, there are often discontinuous lenses 0.5 to 2 ft (150 mm to 600 mm) of soft clay immediately above the rock. In either case, the soft clay supports little or no part of the overburden and foundation loading.

When the stiff residual soil horizon is thicker than the spacing between load concentrations, such as the columns in the structure, the effect of the stress concentrations from these loads is very small compared with the effect of the average load of the structure over its total area. In this situation, settlement will be largely independent of the actual foundation bearing pressure. So long as the foundation pressure does not exceed the safe bearing capacity of the stiff clay, the settlements will be largely controlled by pinnacle punching and the relatively low compressibility of the residual soil.

The project design, therefore, should minimize excavation that removes the stiff residual soil. If there are any solution depressions or sinkholes nearby, the design should provide for correction of those that can be identified plus more detailed exploration to identify hidden erosion domes near the ground surface that have not collapsed. Those also should be corrected.

A structure at the ground surface shields the soil from the infiltration of rainfall and thus reduces the opportunity for the enlargement of existing ravelling erosion domes and the formation of new domes. However, that same structure could also aggravate dome development by adding water from leaking pipes and poorly designed surface water drainage. Therefore, design should emphasize minimizing the opportunity for water infiltration into the residual soil. Large depths of stiff residual soil, small structural loads and projects that do not impound or channel water toward the structure are favorable to foundations supported by the residual soil or on deposited soils above the residual soil.

### 7.3 SWELLING AND SHRINKING OVERBURDEN SOILS

Some residual soils from limestone are highly plastic (i.e., plasticity index exceeding 25): they swell upon wetting and shrink upon drying. Foundations supported on such materials rise and fall with the ground movement. The same design measures are used for foundations over these swelling and shrinking soils derived from limestones as those employed elsewhere over such materials. Usually, there are two courses of action. First, the highly plastic materials are removed to the depth of the seasonal soil moisture changes that generate swelling and shrinking. The soils are replaced with compacted crushed stone or soils that do not swell or shrink appreciably. Second, the foundations are supported below the level of the moisture changes and they are isolated from the surrounding soil that continues to swell and shrink.

If the seasonal moisture changes are severe, the required soil removal or foundation excavation may be so expensive that deep foundations to stiff soils below the swell-shrink zone or to rock may be more economical. Replacement of the swelling and shrinking soil with non-swelling but pervious fill, such as sand or crushed stone, will increase the hydraulic conductivity and thereby aggravate the hazard of future erosion dome formation from surface water percolation. These issues must be resolved before proceeding with foundations on the overburden. For a more detailed discussion of foundations on swelling and shrinking soils, refer to text books on geotechnical engineering or a specialized text such as Chen (1975).

### 7.4 FOUNDATIONS SUPPORTED DIRECTLY ON OVERBURDEN

Column and continuous strip or wall footings and mat or slab foundations usually can be supported on firm to stiff residual soil, filled solution depressions, and plugged, filled sinkholes. Bearing capacity (i.e., the resistance to foundation punching into the ground) is seldom critical in the upper stiff clayey residual soils over limestones or in well-compacted fill as long as the thickness of the stiff residual soil or compacted fill is greater than approximately 3 times the foundation width.

Both bearing capacity and settlement analyses for shallow foundations on the residual soil or compacted fill are the same as for other soil profiles. Basic analysis for bearing capacity can be found in text books on soil mechanics, geotechnical engineering, and foundation engineering (such as Sowers 1979) and will not be discussed further here.

The stiff residual soil is usually over-consolidated and, therefore, has relatively low compressibility. However, settlement can occur by distortion of the residual soil blanket where it is impaled onto narrow rock pinnacles as shown in Fig. 7.1. Additional load provided by new fill or by the overlying structure causes the residual soil to be impaled more deeply. The downward movement of the residual soil mass is resisted by its sliding against the pinnacle. If the pinnacle becomes wider with increasing depth, as is the usual case, the soil-to-rock shear is enhanced by the increasing normal force between the pinnacle and the stiff soil, a form of wedging resistance. The pinnacle punching resistance can be modeled in a finite element analysis in two dimensions employing combinations of rectangular and triangular soil element prisms. A simple two-dimensional model is illustrated in Fig. 7.2. Each element distorts or slides against its neighbor in accordance with its properties of shear strength, rigidity as expressed by its Poission's ratio, and its modulus of elasticity, in response with the changing forces applied to the system. In the model of Fig. 7.2, the soft soil and the stiff clay overburden

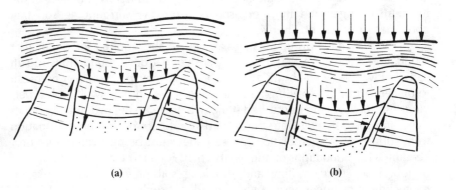

(a)                                                (b)

FIG. 7.1.  Mechanisms of Shear and Compression in the Stiff Clay Overburden on the Pinnacle Surfaces and in the Slots Between Them Accompanying an Increase in Loading at the Ground Surface (also See Fig. 3.6): a. Before Surface Loading; b. After Surface Loading

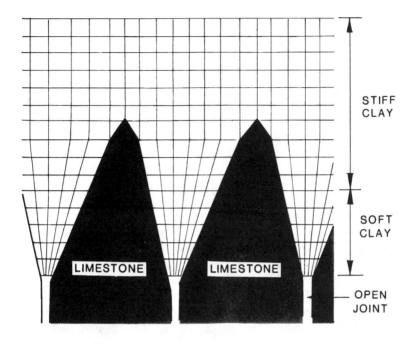

*FIG. 7.2.   Finite Element Mesh Configuration for Analysis of Settlement of the Stiff Clay Overburden by the Punching of the Pinnacles Deeper into the Stiff Clay above Caused by Greatly Increased Ground-surface Load*

have greatly contrasting properties, based on appropriate laboratory tests. The rock pinnacles are usually so rigid compared with the soils that they are relatively immovable. The analysis involves complex iterative solutions to determine the stresses and strains of each element considering the distortion and relative movement between the stiff clay and rock and the volume changes of the soils under the increasing surface load. It permits a prediction of the ground surface settlement when the loading is sufficient to cause the stiff clay to punch downward, and to either compress the soft clay or to extrude it into open voids in the rock below. Analyses by Sams and Moshen (1992) indicate that the compression or resistance to extrusion of the soft clay does not influence the results significantly; the settlement is largely controlled by the distortion of the stiff clay and its sliding downward against the pinnacles. This mode of settlement appears to be significant only when the surface loading is such that it greatly increases the downward force on the stiff clay elements between the pinnacles.

Because the soft zone is so weak and compressible compared with the stiff overburden, the engineering properties of the soft zone seldom influence the amount of settlement. The soft soil either compresses easily, offering little resistance, or extrudes downward through the fissures in the rock into the rock voids below.

In areas where the limestones or their residual soils are blanketed by loose sands, the sand bearing capacity may govern the design of small foundations as it usually does when no limestone is present. In addition, the potential for sand settlement and liquefaction from strong vibrations and earthquakes must be evaluated as for sands in other environments.

In areas where the limestone is blanketed by impervious, non-conforming relatively impervious soil deposits, surface solution of the rock may be very superficial. In such cases, only the areas with widely spaced rock joints and occasional faults may present risks for future solution activity. Such soil and rock profiles are found in areas of glacial plowing followed by deposition of relatively impervious glacial till or glacial lake clays on the plowed limestone surface. In such situations, the soil overburden protects the limestone surface from downward percolating water. However, the areas immediately above continuous deep and wide fissures in the limestones may become sites for very local but severe solution enlargement and sometimes erosion of the soil above. This occurs when the piezometrical level in the groundwater fluctuates above and below the rock surface (a condition of alternating artesian and normal groundwater pressure). Where such areas of fracture can be identified, it is possible to straddle or bridge over the suspect area by battering deep foundations, as shown in Fig. 7.3. Steel H piles that can bite into the rock in order to support the inclined load without sliding have proved practical for such situations. However, any type of deep foundation that can be constructed at an angle or vertical deep foundations with a bridging beam over the enlarged slot could be used.

Excavating on a site deeper than the topsoil zone usually removes some of the stiffest materials in the soil profile. However, this may be partially offset by decreasing initial vertical stress in the stiff soil below the excavation level by removing the soil weight. As long as the increased stress imposed by the foundations of the new structure does not exceed the original stress in the remaining residual soil between and immediately above the pinnacles, the foundation loads are not likely to produce serious settlement. However, if the foundation stresses increase the soil stresses at the pinnacle and slot level, there can be some settlement by pinnacle punchings. Exposure of the deeper parts of the stiff overburden by excavation sometimes aggravates soil dome development below because the downward seepage path is shortened. When the remaining stiff soil is exposed to alternate wetting and drying, vertical shrinkage fissures often develop. High rainfall or

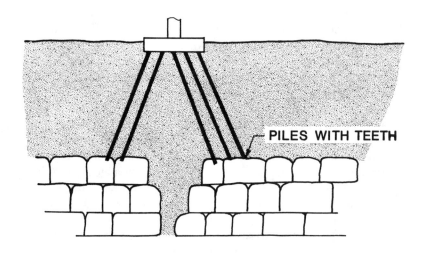

PILES WITH TEETH

*FIG. 7.3.   Local Pile Support Over a Small, Isolated Solution-enlarged Slot or Cavity in the Underlying Rock*

water pounding on this cracked soil can further aggravate soil dome development.

Although all appropriate measures have been taken to identify potential soil overburden defects and the defects have been properly treated, there is always some risk that soil or dome collapse could develop in the future. Some of the risk arises from the design and in the construction of the structure. These include improper surface water drainage and leaking water pipes and sewers. The risk can be increased by changes off-site, such as lowering the groundwater level. However, in most karst areas, the risk is small and catastrophic failure is so infrequent that the public usually ignores it. The ignoring of the risk of sinkhole collapse is demonstrated by the density of population in many areas prone to sinkhole development.

## 7.5 BUILDING FOUNDATIONS RESISTANT TO SOIL COLLAPSE

For structures, such as hospitals, occupied by people who cannot be easily evacuated and which must continue to function despite some catastrophic event, and for power plants, communication systems, police, and fire-fighting facilities, where even a short interruption of service is detrimental to public welfare, it may be prudent to further minimize even a very small risk. In such situations, the structures involved are designed to tolerate a large dome collapse, which is assumed to occur at any critical, random location beneath the structure. For this case, it is reasonable to design for one occurrence at a time because even in very sinkhole-prone regions, simultaneous, closely spaced dropouts are rare. A random location beneath a structure is a reasonable presumption because it is usually impossible to predict where one might occur.

The size of the potential zone of no foundation support is the opening of a hypothetical erosion dome at the time of its collapse (Fig. 7.4). Experience demonstrates that the diameter of the potential loss of support is larger than the diameter of the initial dropout at the ground surface, with its overhanging rim, but considerably less than the ultimate funnel-like opening of the sinkhole after its rim sloughs and caves in following the initial dropout. That width can be estimated from the size of nearby sinkholes underlain by the same limestone formation and with the same thickness of residual soil cover. Where there are no sinkholes, the effective width of an erosion dome collapse can be estimated as being greater than one-half, but less than equal to, the thickness of the residual soil overburden. The stiffer and more clayey the residual soil, the larger the ravelling and erosion dome can become at the time of its collapse. If the soil is cohesionless, it may be narrower than 10 ft (3 m).

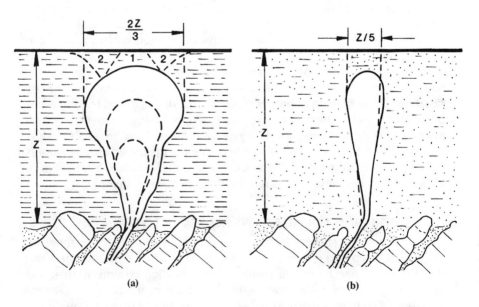

*FIG. 7.4. Potential Sinkhole Diameter. a. Cohesive Soil; b. Low Cohesion Soil*

The sinkhole-resistant structure is designed so it will not fail should there be a sudden catastrophic loss of foundation support because of collapse of such a soil dome. Structural deformation and even non-critical structural damage can be tolerated, provided the overall stability of the structural system is not reduced to the extent that people are injured or the structure's contents cannot be evacuated. Moreover, the function of the remainder of the structure must not be significantly impaired. It is presumed that some rearrangement of people and equipment will be necessary until the potential collapse zone is repaired.

A number of structural systems, including both the foundation and the upper parts of the structure, have been designed to resist the effects of a single dropout. Mat foundations are well suited for local loss of support designs (Fig. 7.5). The mats are thickened and reinforced so they become raft-like. If very large domes and potential dropouts are anticipated, the mat can be combined with several of the floors above to form a hollow, box-like, three-dimensional beam. The outside walls and interior partitions serve as shear transfer diaphragms. Such box foundations have been used in areas of deep compressible soils, such as Mexico City, to resist the distortion of severe differential settlement caused by consolidation of the foundation soils. Similar mat systems can be designed to resist catastrophic failure from a dome collapse.

After construction of the resistant foundation-structure system, it is necessary to monitor the structure regularly for movement and distortion and periodically inspect the structure and the soil beneath it for signs of dome

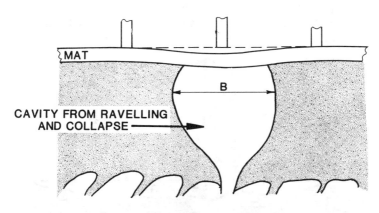

**MAT REINFORCED TO RESIST FAILURE
FROM CAVITY COLLAPSE**

*FIG. 7.5.   Mat Foundation Spanning a Potential Sinkhole*

development. If subsidence is discovered before it becomes a collapse, remedial measures can be undertaken before there is severe distortion in the structure. A simple monitoring system can be installed to permit direct observation of the ground contact with the foundation slab. Small observation holes in the mat are provided to probe the contact of the mat with the soil (Fig. 7.6). Alternatively, sensors can be installed to measure changes in the contact pressure between the mat and the foundation soil. A reduction in that pressure would suggest erosion dome subsidence. Measuring changes in the elevations of points on the mat surface can warn of changes in the support of the underlying soil; however, considerable soil movement is likely to occur before the lack of support is reflected in bending of a stiff mat foundation. Sophisticated electronic sensor systems that measure the structural strains and relate them to the pressures on the soil-structure interface can be devised to monitor and diagnose the effects of changes continuously. Once lack of support is noted, remedial measures can be undertaken before a catastrophe develops. Where such subsidences have been measured in existing structures, grouting, as described is Chapter 6, has been the method chosen to prevent further movement and structural distress.

### 7.6 AN EXAMPLE OF A SETTLEMENT AND COLLAPSE-RESISTANT DESIGN

A large stiffened mat was utilized for the foundation of a multi-story hospital to minimize the risk of building destruction from sinkhole development. The preliminary site investigation disclosed an old, apparently prehistoric sinkhole centered under the proposed building. Moreover, there were several other sinkholes in the vicinity, some of which had developed during

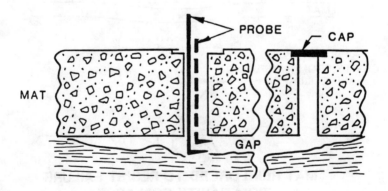

FIG. 7.6. Simple Monitoring Point and Probe for Sensing Loss of Soil Support Beneath a Foundation

the previous century. The geotechnical engineer, architect, and structural engineer recommended to the sponsoring agency that a new site be secured that was less prone to sinkhole subsidence. However, the politicians who had promised construction of the hospital in that neighborhood insisted that the public policy demanded exactly that location. Therefore, appropriate measures should be taken to minimize any risk. Ultimately, the sponsoring agency agreed that it would accept the risk if extraordinary measures were taken during design and construction to minimize possible foundation failure and structural collapse.

The sinkhole infilling was removed to a depth of 50 ft (15 m) and replaced with compacted fill. Carbon 14 dating of wood fragments from the sink hole throat indicated the sinkhole developed approximately 15,000 years ago, probably during a period of sea level fluctuations. The limestone below, at a depth of approximately 80 ft (24 m), contained numerous deep surface pits and slots and a few scattered small cavities. Pile support, therefore, would be questionable, even when strict pile-driving criteria were imposed.

The rock surface was cap-grouted to minimize future soil dome development. In addition, the cavities in the upper rock strata were grouted to minimize groundwater circulation beneath the site. This would further reduce the risk of dome development and improve the rock integrity. The site beyond the building footprint in areas of parking was preloaded to minimize potential settlement from any small, unseen filled depressions that might be present.

A study of new sinkholes in the vicinity found that none exceeded a 30-ft (9-m) diameter after the rims had collapsed. The mat was designed so a 35-ft (10.6-m) wide dropout could occur directly beneath any column in the structure. The mat was made sufficiently strong and rigid to prevent structural failure. However, there might be some visible cracking in the mat concrete in the immediate area of subsidence and non-structural cracking in interior partitions. A program of mat monitoring was begun to warn of any possible movement. After more than 30 years since construction, there has been no sign of distress or settlement; the structure is performing as anticipated.

## 7.7 CORRECTIVE MEASURES

When sinkholes occur in populated areas or in the vicinity of structures, emergency steps are necessary to protect people and property. Although sinkholes can occur anywhere in areas underlain by limestone, they sometimes concentrate near structures, making them especially hazardous. Although structures can inhibit sinkhole development by shielding the ground from rainfall infiltration, they can also aggravate it by improper disposal of surface drainage and waste water. Moreover, the concentrated load of a surface foundation can precipitate failure by increasing the stress on an undetected dome. When sinkholes appear close to structures, their occurrence usually can be linked to the aggravating effect of improperly handled water or failure to follow the precautions discussed in this chapter.

Regardless of the cause, when sinkholes develop, the following measures should be considered: 1) Implement procedures to protect people from injury and minimize property loss in and near the hole; 2) repair the structure, if that is practical; or 3) demolish the structure, fill in the hole, and make the site an asset rather than a hazard or liability.

Because a new sinkhole is a curiosity, it attracts inspection by those who are challenged by the unknown as well as those with a scientific interest. In another sinkhole at the college which lost a lake in a sinkhole (Fig. 4.5), a student inspected the sinkhole throat and a cave adjoining it after midnight when the campus patrol was not very active. His only equipment was a flashlight and a rope. He climbed over the fence that had been erected around the sinkhole and lowered himself into the hole on a rope tied to the fence, and explored the underground opening for 1 hr. Because no one knew of his being there, he could have become a missing person if a collapse had occurred.

The first step in treating a new sinkhole is to place a secure barricade around it or temporarily fill in the bottom of the depression, followed by constructing a fence around the immediate area to keep the public away from danger and from interfering with corrective work. Any overhangs should be removed to prevent rim collapse and further risk to people and property.

If a structure is involved, steps should be taken to remove any contents that are worth saving. At the same time, structural engineers and engineering geologists or geotechnical engineers will determine if the structure should be either repaired or demolished. It is to be demolished, that is done promptly because it remains to be an attractive hazard. (The sinkhole can be repaired as described in Chapter 6.)

If the structure is to be restored in place, the engineering and construction problems of working in an unstable area control the remedial measures. The first step is to prevent further activity in that sinkhole. The throat must be closed despite the obstruction of the work space by any damaged structure. Usually this means filling the throat by techniques that can be implemented from a distance. Ordinary low-pressure sand cement grouting, compaction grouting, and jet grouting have been used depending on the working space, cost, the availability of the equipment, and the nature of the soil. The compaction grouting is better adapted to sandy soils and softer clays. The jet grouting is better adapted to stiffer materials. Ordinary sand-cement grout has been used. However, this provides more chance of a reawakening of the sinkhole because inclusions of soft soil may remain within the grout and become the focus for renewal of deep soil erosion. Once the throat has been closed, the remainder of the hole is filled with as stable a fill as is possible, considering the limited working space available and the intended future use of the site.

Unless a very strong incompressible fill can be placed, it will be better to re-support any structure by *underpinning*. This means constructing new foundations under the structure, preferably supported on the underlying rock. Such foundations are described in Chapter 8. Finally, steps are taken to minimize the site conditions that were responsible for the sinkhole. However, there is no guarantee that a similar event will not take place at the same point or nearby in the future.

# CHAPTER 8

# FOUNDATIONS SUPPORTED ON LIMESTONE

## 8.1 NEAR-SURFACE ROCK SUPPORT

Where the rock is shallow and the rock surface is not seriously disrupted by numerous wide slots and cavities, direct support of foundations on the rock surface is appropriate. If the rock is sound with no cavities close to the rock surface, it is highly unlikely that the rock strength will be a factor of the rock-bearing capacity: the rock strength is likely to exceed the strength of the concrete. If the limestone contains vertical fractures spaced closer than the width of proposed foundations, but is otherwise intact, the ultimate bearing capacity of the rock is approximately equal to its unconfined compressive strength. This strength, divided by an appropriate safety factor, is often far greater than the *allowable load* (i.e., the maximum bearing pressure that most design codes allow for foundations supported directly on rock). When very closely spaced rock discontinuities are present, they limit the foundation design bearing pressure because of the small deflections that can occur as the intact rock fragments move against one another under loading. The fractured rock behaves similar to a densely packed, angular, cohesionless soil.

A level, flat rock surface is not necessary for foundation support, as long as the limestone surface is clean and is no steeper than approximately one-half the angle of friction between the hardened concrete and the rock that supports it. Typically, this means a rock slope that is no steeper than approximately 15 to 20 deg. If the surface is steeper, blasting to make a more level surface should be avoided: it can produce additional surface fissures that weaken the rock. Instead, shear resisting pins can be installed more quickly and without the legal restraints imposed by using explosives. Holes are drilled in the sloping rock surface from 6 to 12 in. (150 to 300 mm) deep. Steel pins, usually pieces of ordinary reinforcing bars, are grouted into the holes. These are sized so that they provide the proper margin of safety against sliding (Fig. 8.1).

When there are wide fissures or well-defined voids beneath the foundation, the foundation gains its support from the remaining intact rock acting as unconfined columns. In order to prevent local spalling of the edge of the fissures under and immediately beneath the foundation, the rock is supported laterally by filling each fissure with concrete to a depth of between 1 and 2 times the fissure width (Fig. 8.2). This is termed *dental filling*. The minimum depth applies to near-vertical fissures that narrow with increasing depth; the greater depth applies to parallel sided fissures, those that enlarge with depth, and to fissures dipping steeper than approximately 45 deg. The dental filling carries little or no load. When the fissure dip is flatter than 45 deg, the entire area of the fissure beneath the foundation should be filled. Alternatively, the overhanging rock is removed below the foundation. It is essential that the rock surface in contact with the concrete be clean in order to minimize any displacement between the two materials upon loading. Depending on the width of the fissure compared to the foundation width, it may be necessary to augment the foundation reinforcing where there is a very wide fissure below it.

When a very wide slot or pit is found beneath a foundation, it is expedient to structurally bridge over that opening with a properly reinforced beam or slab (Fig. 8.3). When the rock surface contains numerous elongated slots that involve many foundations, it may be better to connect all the foundations together in continuous beams or grid-beams in order to bridge over randomly spaced openings of varying widths. In this way, it will not be necessary to design a new foundation for each fissure or pit encountered. However, this does not eliminate the need for dental filling of the fissures to

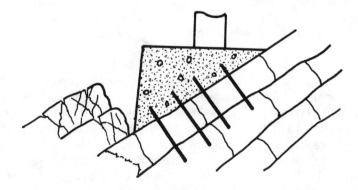

FIG. 8.1.   Shear Pins Grouted into Holes Drilled in the Rock to Resist Sliding of a Foundation on a Steep Rock Surface

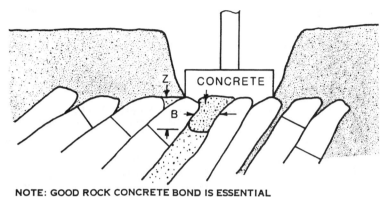

**NOTE: GOOD ROCK CONCRETE BOND IS ESSENTIAL**

*FIG. 8.2. Dental Filling of Wide Slots in Rock Beneath a Foundation Supported Directly on Rock*

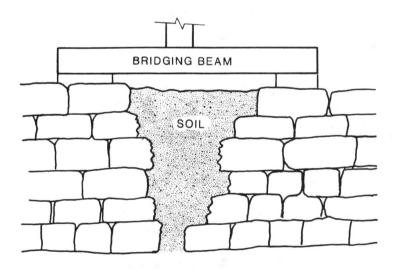

*FIG. 8.3. Bridging Beam Over a Very Wide Soil-filled Slot*

prevent local crumbling or crushing of the rock immediately adjacent to the fissures. Alternatively, if the exploratory data suggest fissures of various width with no distinct pattern in their occurrence, the structural engineer can prepare designs in advance for fissures of varying size and placement under each foundation. The resident engineer or site geotechnical engineer-engineering geologist then selects the appropriate design after the rock and its defects are uncovered during construction. There will be no delay in order to redesign the foundation when each defect is uncovered. This design-before-the-problem is practical when most of the foundations are of similar loaning.

In one project with numerous slots of varying width, a job change order was required followed by a week of delay to redesign the foundation and fabricate reinforcement when each new defect uncovered. The delays were so time-consuming and costly that the foundation engineer pre-designed foundations for the entire range in fissure sizes and the column loads expected at the site. The resident geotechnical engineer then selected which alternative design to use: the only delay was in fabricating the reinforcing steel, i.e., a few hours in most cases.

Dipping, fractured rock may require reinforcing, using grouted dowels or tensioned rock bolts, depending on the degree of restraint needed. The objective is to bind the rock blocks together (Fig. 8.4), so that the rock can support the foundation load without excessive rock movement. Text books of rock mechanics that emphasize fissures, such as Goodman's **Introduction to Rock Mechanics** (1989), can be helpful in selecting bolt patterns, lengths, and tensioning, where needed.

## 8.2 RISK OF CAVERN COLLAPSE IN ROCK

As discussed earlier, cavern collapses in rock do not occur often in terms of the lifetime of engineering projects. The very slow rate of rock removal by dissolution is responsible for the great infrequency of rock collapses. Even

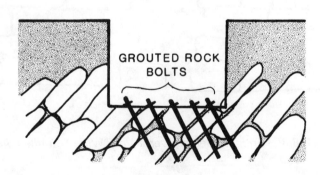

FIG. 8.4. *Enhancing the Mass Stability of Limestone Having Solution-enlarged Close Fractures or Solution-enlarged Primary Porosity Using Grouted Dowels or Rock Bolts*

increasing the flow rates in existing cavities by unlined impoundments does not increase the rate of rock cavity development appreciably. Instead, the increased underseepage of dams on limestone formations is almost entirely the result of erosion of soil filling the cavities. Virtually all cases of significantly accelerated rock solution in rock cavities appear to be related to severe groundwater pollution by acids, irresponsibly disposed of by dumping, released inadvertently from accidents, or produced by the oxidation of sulfides from mine wastes and the excavation of rocks containing iron sulfide (*iron pyrite or fool's gold*). The iron sulfide (and other sulfides) oxidize to form sulfurous and sulfuric acids that dissolve limestones rapidly.

When a heavily loaded foundation is placed directly above a wide cavern with a thin roof, it may cause the roof to fracture and collapse. The roof resistance depends not only on the width of the cavern and the thickness of the rock roof, but also on the fissure patterns and fissure widths, the tensile and the crushing strength of the rock, and the friction between the rock blocks as defined by the irregularities (*asperities*) of the fissures and bedding surfaces. The capacity of such a roof is investigated by the methods of rock mechanics as described by Goodman (1989).

The potential for collapse of the rock roof of a cave, including the remedies possible to prevent it, is evaluated in two stages. As a part of the preliminary site investigation (which includes limestone geology), the site is examined for sinkholes and subsidences. Borings are made to determine the thickness and engineering properties of the soil overburden and the general condition of the limestone at the location of typical sinkholes. This work uses the methods described in Chapter 5. These data are evaluated in terms of their impact on the design of foundations for the proposed structure. If the stress increase at the level of the rock surface resulting from the load concentrations of the proposed structure is smaller than 5% to 10% of the stresses due to the existing soil overburden, it is not likely that the new foundation loads will cause collapse of the roof of a cave if one is present. When this is the case, the remaining geotechnical investigation progresses as necessary for foundation design.

If the foundation stresses are high, if the borings have found shallow caves with thin roofs, or if shallow caves are present nearby, a more detailed investigation is needed to specifically determine the capacity of a cave roof beneath the proposed structures. Experience in rock tunnelling suggests that a cave roof of reasonably sound rock or rock with tightly closed fractures is likely to be safe if the rock thickness is greater than the cave width. If the rock is fractured, the roof thickness should be somewhat greater than half the cave width depending on the fracture pattern and the width of the fractures. The thickness of the rock beneath each load concentration is determined from the investigation data (augmented as needed by percussion drilling at each location). If a shallow cave with a thin roof is found, additional percussion borings can be done to better define the cave limits. The borehole video described in Chapter 5 may help to define the size of the cave. If the borings indicate that the cave is large, and above the groundwater level, a large diameter hole can be made to permit the geotechnical engineers, engineering geologists, and local cave explorers to enter and map the extent of the cave and the structural condition of the cave roof. The stability of the roof of each cave subject to a large load increase from the proposed construction is analyzed by the methods of rock mechanics.

When the rock cover is too thin, there are three alternatives. The first is to shift the structure to avoid the cavern. This will often be the fastest and cheapest answer to the shallow cave problem. The second alternative is to transfer the load of the structure or foundation at risk below the level of the cavern floor. The third is to evaluate the roof capacity using the methods of rock mechanics as stated previously.

One method of transferring the load deeper is to fill the cavern with concrete in the critical location (Fig. 8.5). In order to minimize the concrete volume, the portion of the cavern subjected to stress increase from the foundation loading can be isolated by bulkheads or by placing dams of low slump concrete at the limits of necessary filing. These restrict the flow of concrete beyond the area needing strengthening. Filling in large caverns usually requires access holes in order to build bulkheads to restrict the concrete flow. These also allow a more detailed examination of the rock roof. When there is water flow in the cavern, a concrete plug will disrupt the

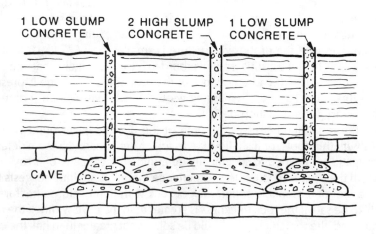

FIG. 8.5.   Local Filling of a Small Cavern Having a Thin Rock Roof Beneath a Very Heavy Ground Surface Load

groundwater regime and possibly aggravate solution or erosion dome development elsewhere. The effect of the filling on the cave environment, including aggravating rock solution and cave development elsewhere, plus the cost of large volumes of concrete, limit the use of this approach.

It is also feasible to extend the foundation through the cave to the cave floor, if it is sound rock. This has the advantage of not seriously disrupting any present or future water flow in the cavern. Piles can be used by drilling through the cavern roof and then driving the pile onto the rock. Corrosion protection is necessary for steel piles. Drilled shaft foundations, discussed later in this chapter, also can be  used. This requires using a cased shaft through the cavern (Fig. 8.6), in the same way casing is used for foundations in areas underlain by old mines. The procedure is the same as for areas of soft soil overburden, as described in Section 8.7.

## 8.3  GREATLY ENLARGED PRIMARY POROSITY

A second hazard for foundations directly supported on rock is limestone that has experienced aggravated solution of its primary porosity or numerous closely spaced fissures sufficient to reduce the limestone's strength or increase its compressibility significantly. Foundation design for such situations is based on laboratory tests of large core samples (at least 4 in. (100 mm) in diameter) so that a representative distribution of intact rock and voids can be evaluated. Both consolidation and triaxial shear tests are needed to estimate the bearing capacity and settlement, in the same way as for soils. If foundations resting directly on the porous rock are not feasible, or economical, one

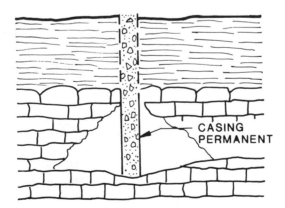

*FIG. 8.6.  Extending a Shaft (or Pile) Foundation Through the Roof of a Cavern with a Thin Roof to Sound Rock in the Cavern Floor*

alternative is deeper foundations, depending on the dissolution profile. A second alternative is improving the rock by grouting, as described in Section 8.4.

## 8.4 ROCK IMPROVEMENT BY GROUTING

When the rock mass is riddled with small cavities, it is usually possible to improve its strength and reduce its compressibility by filling a large proportion of the cavities with portland cement grout. This also reduces the possibility of collapse of small caves, erosion of overburden soil into cavities, and the formation of additional erosion dome cavities in the residual or deposited soil overburden. When the rock cavities are relatively free of soil filling, dramatic improvement is possible. However, despite careful grouting, not all the primary porosity will be filled. The larger open, interconnected cavities will be sufficiently filled so that they will not collapse under load. However, some of the smaller cavities without connections to the cavities being grouted may remain open. When such cavities are partly or entirely filled with soil, the long-term structural integrity of the rock may not be greatly improved by grouting. The immediate potential for the circulation of water, continuing rock solution, and erosion of overburden soil into cavities is reduced, but eventually some of the soil filling will be eroded away. The grout remains as irregular discontinuous bodies in the smaller cavities and gravel- and boulder-size masses littering the floors of the larger cavities. If the primary porosity is so great that it requires filling, a test area should be treated. Large cores of the grouted rock are tested for strength and continuity before a decision is made to rely on this remediation.

Grouting is very expensive both because of the cost of introducing the material into the appropriate cavities through numerous drill holes, and because a great deal of the grout is often wasted by its flow through the larger cavities into areas of no concern at the site (and sometimes off the site into property owned by others). Grouting also disrupts the groundwater flow in the cavities, forcing the water to seek new paths. This can cause increased overburden ravelling, erosion dome formation, and dome collapse elsewhere; sometimes this also occurs on property owned by others. Such trespass on adjacent sites has ethical, legal, and financial liability implications that must be considered.

It is difficult to estimate the amount of grout to fill the open cavities beneath a project. The author has utilized a simple procedure that has been reasonably successful. It requires a large number of continuous rock cores that define the size and shape of the voids, supplemented by carefully monitored percussion drill records that depict the lengths of continuous rock, as well as the lengths of rod drop indicating voids, and slow rod drops or low drilling resistance indicating soil filling. The percentage of open or water-filled cavities in each boring between the elevations to be grouted is determined from measurements of the continuous rock cores supplemented by the percussion drill rates of penetration. A weighted average of the percentage of open cavities, within the rock mass (and overburden) beneath the structure, and within the elevations to be grouted, is computed. This percentage, multiplied by the total soil and rock volume between the elevations that requires improvement, is the approximate volume of voids to be grouted. Based on experience, the real amount of grout is likely to be from 0.7 to 2 times as much as the volume computed. It can be more than that computed because some of the grout is lost in cavities beyond the specified levels and limits; it can be less because some of the cavities inferred from the cores or penetration rates of percussion drilling may not be interconnected or filled with soil. Grouting voids in limestone requires the largest practical diameter of grout holes and the largest proportion of aggregate particles in the grout mix consistent with pumping the grout into the "diameters" of cavities to be filled. Eighty-five percent of the aggregate should be smaller than one-half the size of the smallest opening that is required to be filled. The proportion of inert solids in the grout is maximized to reduce grout shrinkage, consistent with nominal strength and the ability of the grout to flow readily. Water-tight integrity is not critical, as it is for dam foundations, because the water pressure differentials are usually small. High strength is usually not necessary because the chief objective of the grouting is to minimize the potential for cavity collapse and to reduce, but not necessarily completely prevent, the groundwater circulation.

Grout mixes are developed by trial and error. Typical grouts include portland cement, fly ash (if available), sand, and, often, fine gravel. Partially rounded aggregates are preferable to very angular for better pumping; however, pumpability and flow can be improved by adding fly ash. A small percentage of bentonite (up to approximately 5%) or chemical admixtures can also improve pumpability. Reasonable strengths are approximately 500 psi (3.5 MPa) to 1,500 psi (11 MPa).

Grouting holes are drilled in a two-dimensional grid pattern, spaced 10 to 30 ft (3 to 10 m) apart, depending on the cavity sizes and their likely interconnection as inferred from the site investigation data. Auger holes through soil and soft rock and percussion drill holes in rock are usually adequate for grouting. In either case, the depths and sizes of cavities found in the grout borings should be recorded in order to plan for additional holes, if necessary. The grout holes are drilled at an angle, if necessary, to intersect the greatest number of fissures; vertical grout holes are not likely to intercept a large proportion of the vertical fissures that dominate level bedded rock.

The holes around the perimeter of the area to be treated are drilled first. In the area remaining, alternate holes designated as *primary holes* are drilled and grouted next. If the amount of grout required to build up the specified grout pressure is less than a quarter of that estimated based on the exploratory data, the program should be re-evaluated; the benefit of grouting may be limited. If the volume is within −50% to +50% of that anticipated, then proceed to grout the remaining or *secondary holes* between the primary. If the secondary holes accept grout at rates exceeding approximately one-half of the amounts of the primary holes, additional, *tertiary holes* drilled and grouted between those holes that took the larger amounts are prudent. Additional grouting between tertiary holes is sometimes necessary in areas of more developed dissolution.

The grout pressure (measured at the top of the hole) should be great enough to break through small soil blockages in the cavities, but not enough to cause heave or hydrofracturing in the rock or soil overburden. The typical maximum grout pressure (as measured at the top of the grouted layer) is equal to the vertical stress caused by the weight of the soil and rock overburden at the level the grout is penetrating into the rock voids. However, somewhat larger pressures may be helpful in breaking through isolated clay deposits that may block some cavities. Such pressures must be utilized cautiously to avoid *hydraulic fracturing* of the rock accompanied by ground surface heave. This occurs when the pressure of the grout in the ground exceeds the vertical stress in the rock due to the weight of the soil and rock above. During grout pumping, the grout pressure and rate of flow are monitored continuously. At the same time, elevations of the ground surface in the vicinity of the grouting are observed. A grout pressure increase followed by a sudden decrease usually means a soil obstruction in a void has been penetrated. In that case, pumping may be continued but at a somewhat slower rate after verifying that there has been no ground heave. When the pressure reaches the overburden pressure or when at least 1.5 times the grout volume estimated to fill the voids in that location have been pumped, that hole is considered to be complete. Where there are large voids shown by the boring data or when large grout volumes are accepted by the rock with little resistance, it is possible that either an unusually large void or a connection to a void leading off-site has been encountered. When this occurs, the grouting is stopped until the grout can develop an initial set. Grouting may then be resumed in that hole, using a more viscous grout if possible. Eventually, a decision must be made whether the large additional amount of grout is of value; at that point, grouting is usually stopped.

When the grout pressure rises close to or exceeds the estimated hydrofracture level and then suddenly drops, the rock may be cracking from the pressure. The pumping rate should be reduced greatly and the ground surface checked for heave. If no heave has occurred, the grouting is continued slowly. A second pressure drop approaching or slightly exceeding the hydrofracture level, particularly accompanied by ground heave, confirms hydrofracture and that hole is considered complete. If there is no further pressure drop or heave, the grouting is continued until the required pressure is reached. When grouting is stopped because of heave or hydrofracture, grouting may be continued in other grout holes. Usually, additional holes are added in areas of heave to fill any gaps that the heave has caused.

When the zone to be grouted is thicker than approximately 30 ft (9 m), the grouting is often done in multiple levels or *stages* of 20 to 40 ft (6 to 12 m) of hole. This permits higher pressures to be used in the deeper parts of the hole without danger of hydrofracturing. One way is to drill to the bottom of the upper stage and grout it. When the grout has hardened, drill deeper in the same hole to the bottom of the second stage and repeat the grouting. However, this is time-consuming because it requires multiple movements of the drilling rigs from one hole to another. An alternative procedure is to drill the entire depth to be grouted. Each stage is successively isolated, from the bottom up, using a temporary, air-inflated block in the hole, termed a *packer*. If there is a tendency for hydraulic fracturing, this procedure permits the highest safe pressure at each level.

Grouting to enhance structural integrity is usually expensive, compared to its tangible benefits. It may be necessary when very heavily loaded foundations are involved such as bridge foundations or very tall buildings. It is not a cure-all for limestone cavity problems. It is most cost-effective where the amount of soil in the cavities is small and the foundation loads are high.

### 8.5 DEEP FOUNDATIONS ON ROCK

When the overburden soil is thick, but not strong enough or sufficiently incompressible for supporting the anticipated loads safely and with tolerable settlement, when the risk of soil ravelling dome collapse is too great, or when the upper surface of the rock cannot support the foundation loading, *deep foundations* supported directly on competent rock may be the answer. These include *piles* (i.e., prefabricated columns that are hammered or otherwise driven through the soil) and *drilled shafts* (also known as *drilled piers* or *caissons*) that are excavated through the overburden and weak rock to sufficiently sound rock. Steel pipe piles smaller than approximately 6 in. (150 mm) in diameter are also known as *pin piles or micro piles*. These are usually filled with cement grout for corrosion resistance and greater stiffness.

There are some ambiguities in deep foundation terminology. Small drilled shaft foundations to rock, in which the concrete is injected under nominal pressure, are sometimes termed *augered piles, mixed-in-place concrete piles,* or *micro caissons*. These sometimes incorporate steel casing through the soil overburden and often a steel reinforcing bar. If the diameter is smaller than approximately 6 in. (150 mm), they may be reinforced by a deformed steel bar, and pressure-grouted at the bottom into the deeper soil or rock. Some contractors refer to these as *soil nails*. However, the term "nail" implies driving them into the ground by a hammer. A more appropriate term is *dowel or grouted pin pile*.

Although deep foundations often appear to be an answer to all the limestone problems, the installation of these foundations involves new problems and different risks. In addition, it is more difficult (and often impossible) to inspect the critical parts of these foundations. The critical parts include: 1) that part of the foundation in contact with the rock, 2) the rock adjacent to and immediately below the foundation (the zone of transfer of the foundation load to the rock), and 3) the condition of the deeper rock in the zone of significant stress increase from the foundation loads.

### 8.6  PILE FOUNDATIONS

Nearly all of the various types of piles that are in use today have been used in limestone terrain. However, there are special requirements that restrict the adaptability of many types. Piles derive their support by both a combination of *skin friction (side shear)* along the pile shaft above the rock and by *end bearing* on rock below the pile tip. Because the principal reason for using deep foundations in limestone terrains is that the soil overburden (and sometimes the upper part of the rock) is suspect or unsuitable, it usually makes little sense to use types of piles that are designed to transfer a large proportion of their load into the soil overburden by skin friction. These include wood piles, many forms of cast-in-place concrete piles, and tapered steel piles. Because the depth to the level of sound rock is likely to be highly variable, the pile selected must be readily adapted to both shortening (i.e., cutting off if the one being driven is too long) and to lengthening or splicing if it is too short. For example, precast concrete piles are troublesome to cut off; splicing them to make them longer is usually expensive and delays resuming the pile driving. By way of contrast, steel H-piles and steel pipe piles can be readily cut off by burning or spliced by welding with only short delays during driving.

Piles that are driven onto a sloping rock surface often slide downward, either bending or breaking (Fig. 8.7). Sometimes they curl up or buckle. A pile tip that bites into the limestone minimizes such sliding. When the limestone is soft and porous, a narrow or pointed tip can damage the rock, however. An irregular, steeply sloping hard rock surface sometimes will cripple the pile tip, as is shown in Fig. 8.8, unless it is reinforced or otherwise armored to resist damage.

The overall pile foundation system is usually designed conservatively when the piles are driven onto limestones. When there is a well-developed slot and pinnacle limestone surface, the pile lengths vary greatly, even in a single pile cap or foundation. In this case, all the piles in the group will not

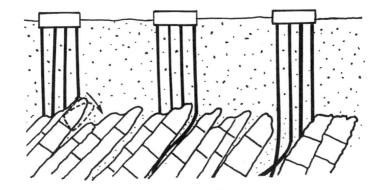

FIG. 8.7. Hazards to Piles Driven Through Overburden to End Bearing on Limestone Below

FIG. 8.8. Crippling of the Tip of a Pile Driven Onto a Dipping Limestone Surface

carry the same load. Instead, the longer piles in each group carry less load than the shorter piles. For example, a pile half the length of a longer pile will support twice the load of the longer pile because the elastic strain in the shorter pile is twice that of the longer pile, as long as the pile cap remains level and the rock at the pile tips is unyielding. The settlement of pile foundations driven to rock is largely that of the elastic shortening of the piles. In areas where the rock surface is highly irregular, those foundations with the longest piles will settle the most. Such differential settlement is ordinarily small. However, when there are great pile length differences, both the rock integrity and the possible effects of differential settlement should be evaluated.

It is also possible that some piles in a group may not support any appreciable load. For example, in driving piles through approximately 60 ft (18 m) of overburden to rock in Florida, it was found that an occasional pile drove as deep as 100 ft (30 m) and one pile was driven 700 ft (210 m) before it reached the required driving resistance. Despite meeting the specified resistance, the 700-ft (210-m) pile had little capacity as part of a group supporting the same column. At best, considering only elastic shortening, it would support only 60/700 (8.6%) of the load of each of the 60-ft (18-m) long piles.

To some extent, the driving of each pile tests the local strength and crushing resistance (*dynamic end bearing*) of the rock beneath the pile tip because each hammer blow is transferred to the rock by an elastic stress wave that travels down the pile. If the rock fractures, the pile penetrates further until the rock is sufficiently strong and continuous to withstand the impact. However, during driving high-capacity piles, sometimes there is continuing slow penetration of the pile under repeated hammer blows. It is difficult to determine whether this reflects seating the pile into the rock surface, crushing of the pile tip, or progressive fracture of the pile shaft. A dynamic pile driving analysis that measures the wave propagation from each hammer blow can help determine what is happening underground. In situations of serious doubt, however, it may be necessary to pull out a pile that is suspected of being damaged by driving. The crippled pile shown in Fig. 8.8 was pulled out because its resistance to driving indicated that something was wrong.

When high-capacity piles are required, it is necessary for the foundation engineer to select a pile strong enough that it can withstand hard hammering when it is necessary and a pile hammer that is light enough that it will not damage the pile when the pile tip reaches solid unyielding rock. At the same time, the hammer must be heavy enough to overcome the inertia provided by the weight of the pile and its cap and the accumulated soil friction on the shaft. The pile-driving criteria should not only specify a minimum driving resistance to obtain the needed bearing capacity but also a maximum resistance that limits the driving when the pile reaches rock in order to minimize rock and pile damage. The dynamic pile-driving analyzer will help determine if a change of hammers is needed for efficient, safe driving.

Where there are thin rock layers underlain by soft soil or a cavity, it will be necessary to penetrate the thin rock. This can often be done by unusually hard driving of the first piles in a group. Otherwise, with each successive closely spaced pile, the weak layer is progressively fractured and eventually broken by rock fatigue. The first piles of each group that did not break through the weak layer are eventually supported by rock that was broken by the subsequent piles. Under the foundation load, those piles that did not initially break the rock will subside more than those that broke through during driving. It is also possible that none in the group will break through during driving. However, the group may eventually do as under their combined final loads when the structure is complete. This can cause severe differential settlement. Good information on the subsurface conditions and experienced engineering inspection of the pile driving are necessary to determine the best approach in such situations. Whenever the site conditions or the pile length variations suggests that the first piles of a group are hanging up on a thin rock seam underlain by softer materials, it is prudent to attempt to redrive the suspect piles. If redriving using the specified driving criteria causes significant added penetration of the first piles driven in a group, all the piles in the group should be redriven. Of course, care must be taken that the redriving does not fracture or cripple the tips of the first piles driven. Fortunately, such problems are not common but they do occur, as the following example illustrates.

During the construction of an addition to a county administration building in Florida, a shallow cavern system was uncovered that extended under the existing building. The existing building was supported by steel piles driven to refusal on rock. Despite the rock bearing, there were some settlement problems, although they were not serious enough to justify investigation. The old driving records showed that some of the piles were approximately 12 ft (3.6 m) longer than the others. The cavern, found a decade later, provided the answer to this anomaly. The longer piles penetrated through the cavern roof and were supported on the cavern floor. The other piles in the same group were much shorter, as they were supported on a hard seam within the thin cave roof. Fissures in the roof beneath the shorter piles suggest that they did not have the required long-term load capacity and that the longer piles were carrying more than their share of the load. Fortunately, there was no severe differential settlement in this situation because of this construction error; the original design was very conservative.

## 8.7 DRILLED SHAFT FOUNDATIONS

The drilled shaft is the modern version of the ancient "well foundation" that began with a hand-excavated hole to rock. Centuries ago, a stone masonry pedestal was built in the wellhole to form a deep foundation resting directly on the rock. Today, a similar hole is excavated by a drilling machine utilizing a variety of drilling tools, from augers to core barrels or batteries of impact hammers combined in one drill head. Diameters from 6 in. (150 mm) to 10 ft (3 m) or more can be drilled deeper than 100 ft (30 m). If the overburden soil is firm and the water table deep, the hole often can be excavated in a few hours to depths of 100 ft (30 m), provided the drilling machine has sufficient torque and a sufficiently long telescoping drill stem, or *Kelly Bar*. In firm clay above the groundwater level, the hole may stand open for hours without support. However, it is rarely prudent to complete the foundation without inserting a temporary supporting tube or *casing* or filling the hole with a *drilling fluid* (sometimes called *drillers mud*) to prevent local caving of the soil in the walls of the hole until the shaft is filled with concrete. If workers or inspectors are to enter the hole, it is imperative that the walls be supported by casing for their safety.

In softer soils, particularly below the groundwater table, it is necessary to insert casing (or utilize drilling fluid as will be discussed later) to support the vertical soil face exposed by drilling. Sometimes the casing will slowly sink by its own weight during drilling; however, it is often necessary to twist it while applying a vertical force to it or to hammer it into the ground. When the casing cannot be forced deeper, it is left in place until concreting. Drilling is resumed with a smaller drill inside the first casing. The deeper hole wall is then supported by a smaller casing or *subcasing* inserted into it. When the shafts are deep and the soil is very loose sand or soft clay, several subcasings may be required. This decreases the shaft diameter materially. Therefore, it is important for the contractor to review the soil and groundwater data before beginning each hole so as to plan for sufficient diameter of the shaft at the bottom. If the hole diameter is reduced too much, the loss of the cross-sectional area can be offset by adding reinforcing steel to increase the shaft load capacity. It may be necessary to enlarge the shaft bottom or rely on drilling into the rock to overcome any deficiency of end bearing from the smaller diameter, as will be discussed later. All casings are removed during concreting as described in Section 8.9, unless the casing strength is included in the shaft capacity (or the casing is stuck so tightly that it cannot be lifted).

Loose material in the hole bottom is removed by various mechanical clean-out devices before concreting. Otherwise, the loose material will consolidate or extrude when the shaft is loaded. However, this may be difficult if the soil overburden or a weathered rock surface produces coarse fragments that are not removed by the drill-mounted cleaning tools. In such cases, it is necessary to inspect the rock surface for thin overhangs, loose rock, and badly fractured or weathered materials in the rock surface. This requires that inspectors and workmen enter the hole, inspect the bottom, and take corrective action, including removing overhangs and breaking boulder-sized loose rock using pneumatic hammers and small explosive charges. These fragments are then readily removed.

Light blasting is sometimes used to level the hole bottom or to deepen it. If too great an explosive charge is used, the rock can be so badly fractured that it will be necessary to make the hole even deeper. However, a level rock surface is not necessary, although it is sometimes specified. Instead, a rough but sound, clean rock surface will resist sliding by shear when the surface is not level. If the rock surface is steeper than half the angle of friction between rock and concrete, 15 to 20 deg., steel dowel pins can be set in the hole bottom in the same way as pictured in Fig. 8.1. Probably the best shaft bottom is bowl-shaped, which aids the hand cleaning and provides resistance to lateral movement, if needed.

When the bottom is below the groundwater level, the hole must be pumped out to permit hand cleaning and inspection if necessary, with the attendant hazard of blowouts from solution fissures filled with soft soil. However, with proper precautions and careful supervision, the work usually, can be done safely. If the engineer or contractor decides against downhole cleaning and inspection, or if the hole cannot be dewatered without danger of blowouts, the hole must be completed as drilled. In this case, the level of confidence may be lower, which must be offset by a larger margin of safety in design.

In badly solutioned or deeply fractured limestone formations, it is essential that the foundation rock be examined for signs of weaknesses and explored below the bottom of each foundation for defects. These include cavities below the rock surface that could allow the rock to crush or break under the future foundation load or clay-filled seams that would allow the foundation to subside as the clay consolidates or extrudes outward under the concentrated foundation load.

One approach to this detailed examination is to make one or more borings at each foundation location before beginning the foundation drilling. If bad rock conditions are found during the early stages of site investigation, this more detailed exploration can be added to the original site investigation program. Although this is the best time to learn of potentially serious defects, the usual exploration time and money budgets seldom allow for this. Moreover, at the exploration stage, the exact foundation locations and sizes are seldom known. Therefore, such detailed work is most often delayed until construction and often until after the foundation hole is drilled. At this time, borings can be made from the ground surface, using a stiff casing or from the hole bottom using hand-held drills. Both diamond-core drilling and percussion drilling are useful. Because the limestone weathering is by solution, it is not always necessary to utilize core drilling for such exploration. The change from soil to rock and back to rock can be determined by percussion drilling almost as well as by core drilling. However, the core drilling provides a more detailed picture of rock fracture patterns that influence rock stability and bearing capacity but at a much greater cost and greater time delay. Typically, the rock below the shaft bottom is explored to a depth of 1.5 to 3 times the shaft diameter, depending on the rock fracture patterns and the foundation loading. The walls of each test hole can be probed to find any small open seams or cavities that would compress or allow the rock to fracture under load transferred by the shaft. This is done with a steel tube or rod measuring approximately 3/8 to 3/4 in. (10 to 20 mm) in diameter fitted with a small, horizontal wedge-shaped tip that is pressed against the hole wall as it is lowered into, or pulled out of, the probe hole. The tip can find open seams less than 0.04 in. (1 mm) thick and clay seams as thin as 0.04 to 0.08 in. (1 to 2 mm). This is illustrated in Fig. 8.9.

If the rock is of insufficient quality to provide the required bearing, some engineers advocate enlarging, or *belling,* the end of the shaft to reduce the end bearing pressure (Fig. 8.10). A reaming tool replaces the drill if the bell

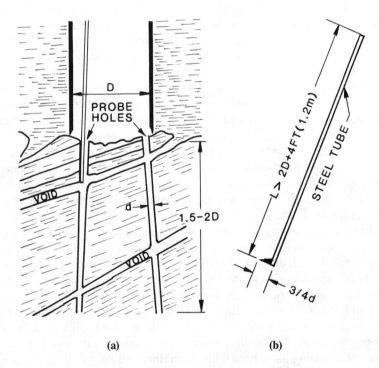

(a)                                    (b)

*FIG. 8.9.   Finding Defects in the Rock Below a Drilled-Shaft Foundation with the Aid of Small Diameter Drill Holes and a Hand Probe: a. Test Holes Drilled in the Rock in the Bottom of a  Drilled Shaft to Identify Enlarged Rock Cracks; b. Feeler or Probe for Finding Rock Cracks in the Walls of Test Holes*

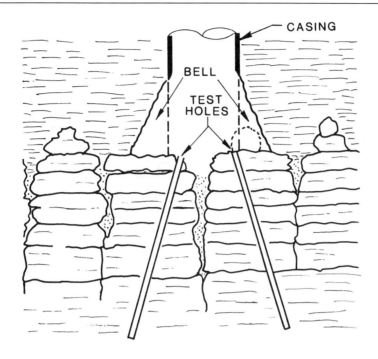

*FIG. 8.10.   Widening or Belling the Base of a Drilled Shaft in Firm Overbur-
den Soil. Test Holes in the Rock Help Identify Open or Mud-filled Cracks
that May Squeeze Shut and Settle Under the Foundation Loading*

is excavated in firm soil above the rock surface or cut into the rock surface,
if the rock is sufficiently soft to permit belling. If a reaming bit is not
available, the bell can be excavated by downhole percussion hammers.
Unfortunately, the soil immediately above the rock surface is sometimes too
soft to stand up within the belled zone. When this is the case and greater
bearing area is required, the foundation must be extended into the rock.
Based on the author's experience and confirmed by many load tests and cost
studies, it is far more effective to extend the shaft deeper than to enlarge the
hole bottom in rock. This has two benefits. First, the deeper rock is likely to
be stronger and with fewer defects. Second, the foundation load is partially
transferred to the rock in the side of the hole by shear, as illustrated in Fig.
8.11a. If the rock surface in the shaft extension is rough and clean, the shaft
concrete bonds to it, and the full shear strength of the concrete provides load
transfer. The effect is similar to that of adding a bell to the bottom of the shaft.
At the level of the base of the deepened shaft, the load is distributed through
the deeper rock in about the same manner as if the shaft had been belled
(Fig. 8.11b).

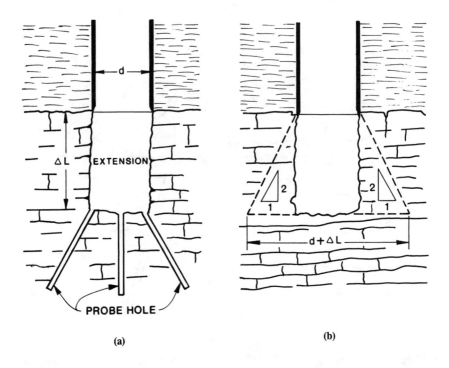

(a)                                        (b)

FIG. 8.11.   Extending the Length of a Drilled-Shaft Foundation to Reduce
the Maximum Vertical Stress in the Underlying Rock by Transferring Much
of the Load From the Shaft to the Rock by Shear: a. The Shaft Extension Into
Intact Rock Transfers Load by Shear and Bond with the Rock as Well as by
End Bearing; b. Computing the Average Vertical Rock Stress Increase by
Assuming that the Total Shaft Load is Distributed into the Rock Below the
Shaft Bottom as if the Embedded Shaft and Surrounding Rock Form an
Inverted Truncated Conical Pedestal

A variety of rock drills are available, from cylindrical cutters set with
tungsten carbide teeth to pneumatic downhole impact drills mounted in
groups to drill holes much larger in diameter than the individual drill units.
Such rock drilling is slower and more expensive than drilling through soil;
however, it is more effective mechanically, as well as cheaper than belling.

A very deep slot directly under the bottom of the shaft (Fig. 8.12) is one
of the more aggravating enigmas in developing the maximum rock capacity
during construction. The usual conservative approach is to drill deeper
because many slots become narrower with increasing depth. However,

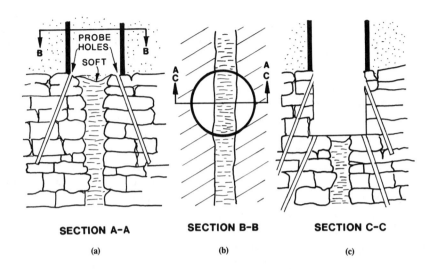

| SECTION A-A | SECTION B-B | SECTION C-C |
|:---:|:---:|:---:|
| (a) | (b) | (c) |

*FIG. 8.12. Remedial Measures for a Shaft with a Deep, Near-vertical Slot in the Rock Beneath It: a. When the Shaft Reaches Rock Probe Holes are Made in the Adjoining Rock to Determine the Rock Continuity; b. Plan View of a Vertical Slot Below a Shaft; c. Shaft Deepened with Added, Deeper Probe Holes (Reinforcing Bars Can be Grouted into the Probe Holes)*

some do not narrow appreciably and a few even become wider with increasing depth.

When the rock on both sides of the slot is intact and sound (as indicated by inspecting shallow impact drill holes (Fig. 8.12c)), it is possible to transfer the entire shaft load to the rock exposed on the walls of the hole by shear (adhesion plus friction). The transfer depends on the roughness and cleanliness of the rock face. The results of pull-out tests and load cells installed at the bottom of the shaft, such as the Osterberg Load Cell (Osterberg 1989), show that shear and the resulting load transfer is usually limited only by the shear strength of the shaft concrete. However, if there is a question of load transfer from the concrete to the shaft, the capacity can be enhanced by setting steel bars in the probe holes to act as shear transfer dowels. If the slot narrows only slightly with increasing depth, as shown by probing into the slot, heavy H-piles driven into the slot can augment the shaft capacity, if necessary.

When there is a large proportion of slots and cavities below the practical limit of depth for drilling the full diameter of the shaft, the ability of the rock to support the concentrated shaft load (including shear transfer) may be questionable. In this situation, *micro caissons* have been drilled below the shaft bottom as an alternative to continuing the full shaft downward by means of extraordinary measures, such as hand mining or special drills. (These are true *caissons* because their steel casings are intentionally left in place and become load-supporting members of the ultimate foundation. They are drilled from the bottom of the main shaft using downhole impact drills inside the micro-casing as described by Chan (1986).) Each micro caisson is battered slightly outward in a different direction so as to engage a larger part of the rock mass below (Fig. 8.13). Such micro caissons have been extended 100 ft (30 m) deeper than the full-diameter shaft. The steel casing of each micro caisson is continuous from the ground surface down through the main caisson to act as a guide for the micro-caisson drilling. The rock cuttings are removed through this casing at the ground surface. Steel reinforcing bars, as necessary, are then inserted into the drilled hole through the casing and the micro caissons are concreted using fine aggregate concrete appropriate to the size of their shafts. Sizes are dependent on the reinforcing required and the size of drills available. Typical micro caisson diameters are from 6 to 12 in. (150 mm to 300 mm).

### 8.8 SHAFTS BELOW THE GROUNDWATER LEVEL

When the rock surface is significantly below the groundwater level, the procedures previously described often must be changed. If the rock slots and cavities encountered are full of firm clay and the thickness of the soft zone

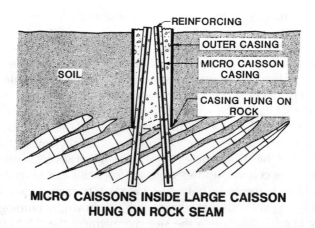

**MICRO CAISSONS INSIDE LARGE CAISSON
HUNG ON ROCK SEAM**

*FIG. 8.13.   Extending the Load Transfer from a Shaft to Micro Caissons Drilled Through Deep Solutioned Rock to Better Rock Below*

above the rock surface is limited, it is usually feasible to lower the ground-water by pumping from within the shaft casing after drilling is finished. This permits hand inspection, hand cleaning (when necessary), and concreting under relatively dry conditions. Alternatively, nearby pumped wells drilled into the soft zone between pinnacles are sometimes effective in lowering the water table. It may be necessary to caulk a few open seams with rags and similar materials to reduce water inflow into the shaft bottom.

If the water inflow is large, there are several alternatives, all requiring experience and all involving risks. Groundwater lowering from deep wells or sumps, including (temporarily) one of the previously drilled but un-concreted shafts, may be effective. As was previously discussed, groundwa-ter lowering generates downward seepage that is often accompanied by ravelling, erosion, and new or renewed dome formation. These events may be followed by ground surface subsidences and new sinkholes in the vicin-ity. While such movements may not affect the structure that is being con-structed (because it will be supported on foundations extending to rock), lowering the groundwater level significantly can endanger neighboring build-ings on shallow foundations, underground utilities, roads, and other near-surface supported structures as the following example illustrates.

Drilled shaft construction, accompanied by dewatering, at a hospital in southwestern Virginia, caused settlement of existing hospital buildings within 150 ft (45 m) of the new foundations. The repairs cost nearly $1 million more than the contract price of the foundations being constructed. The first shaft drilled encountered numerous large, clay-filled cavities in the rock. While drilling deeper to find sound rock, a clay-filled cavity with "artesian" ground-water pressure was encountered 20 ft (6 m) below the top of the rock. This caused mud boils in the shaft bottom. Drilling a number of small diameter test holes in the bottom of the shaft confirmed that the area of high ground-water pressure was extensive. The shaft became flooded with muddy water and pieces of soft clay when attempts were made to extend the shaft still deeper. While the contractor was considering the alternatives, he drilled several of the adjacent foundation shafts. These encountered better rock with limited water inflow at higher levels but somewhat larger flows at greater depths, well below the "artesian" piezometric level. In order to control the water in these new shafts, the contractor utilized the first shaft, still uncom-pleted, as a giant sump. He installed several large pumps in it that were able to lower the groundwater pressure level in the nearby shafts sufficiently to permit completing them without other dewatering measures. The drainage effort was successful. However, large amounts of clay in suspension, as well as gravel-sized clay chunks, were pumped out of the first shaft (now the drain sump) during the completion of the others. Approximately 1 week after the pumping commenced, adjacent existing hospital buildings, supported by spread footings, settled up to 6 in. (150 mm). The settlements occurred as far as 150 ft (45 m) from the shaft being used as a sump. The pumping eroded clay from the rock cavities, from slots between rock pinnacles, and from soft clay zone above part of the rock as was shown by subsequent borings. Eventually, the first shaft that had been used as a sump was completed without dewatering using the tremie process, as will be described in Section 8.9. There was no further settlement. In retrospect, it would have been better to drill and concrete the holes on this site using the wet method as described in the next section.

When the rock defects are below the groundwater level and are large, open, and interconnected, it may not be practical to drain a drilled shaft. Two alternatives are possible. The first is to plug the intersection of the shaft with the adjoining lateral cavities using low-strength concrete to minimize the water inflow. The coarse aggregate for the concrete is no larger than approximately one-fourth the cavity opening to be filled and preferably made from soft rock, as it will be easier to drill out later. The shaft is first allowed to flood. The concrete is then *tremied* into the drilled hole below the casing bottom after the groundwater has reached its equilibrium level (as discussed in Section 8.10). Because the concrete density is greater than that of the water, it will penetrate into the open fissures and cavities in the wall of the hole, displacing very soft clay and water. However, if the water pressure in the hole has not reached equilibrium, the inflowing water may prevent the heavier concrete from extruding into the cavities. When the fill concrete has hardened, the hole is redrilled through the fill concrete in the shaft using rock drilling procedures to extend the hole to a sound foundation level. The redrilling is done with a somewhat smaller drill so as to leave a thin-walled cylinder of hardened concrete to line the hole and prevent inflow more effectively. After pumping out the water, the foundation is then cleaned, drilled deeper, if necessary, and then concreted in a relatively dry hole. The loss of shaft cross-section is offset by adding reinforcing steel to the shaft as needed. This is a last resort method for obtaining a dry hole and is time-consuming and expensive.

### 8.9 WET SHAFTS

Instead of dewatering when rock is below the groundwater level, it is usually expedient to drill, clean the hole, and place the concrete shaft without dewatering. When the casing rests directly on reasonably level, sound rock, it is possible to remove the accumulated soil and rock fragments from the hole bottom by water jetting accompanied by sucking water and solids into submerged bailing buckets with bottom inlet valves or by *air lift* pumping. Alternatively, the hole is kept filled with *drilling fluid*. Drilling fluid is typically a suspension of highly plastic clay in water that increases the fluid viscosity greatly and its density slightly. Occasionally, on-site overburden clay is suitable. However, for ease in mixing and for uniform properties, processed clays, ranging from commercial bentonite to proprietary mixtures consisting mostly of montmorillonite or smectite clays, are used. Various polymer suspensions are also used. Although they are usually more expensive than the bentonite-based mixes, some have the advantage that their viscosity reverts to that similar to water, which can be disposed of more easily. The finer drill cuttings are suspended in the viscous fluid and lifted to the surface by suction bailing or air-lift pumping at intervals in the drilling process, methods most widely used in the United States. In Europe and Asia, the *reverse circulation drilling* is often employed in which the fluid flows down the shaft between the casing wall and the special large diameter drill stem (Fig. 8.14). The drill fluid is sucked out and up through the drill stem. The fluid velocity upward in the drill stem is much greater than the downward velocity between the casing and the drill stem; the upward velocity and the fluid viscosity must be sufficient to carry rock fragments upward from the hole bottom. This requires close control of the viscosity and density of the fluid. Alternatively, the fluid with the drill cuttings suspended

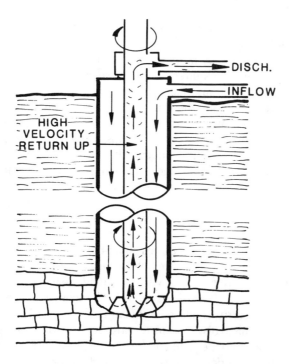

*FIG. 8.14.   Reverse Circulation Drilling to Recover Drill Cuttings*

in it is removed by air-lift pumps through riser pipes between the casing and the drill stem, or by bottom suction bailers.

After cleaning, the hole is concreted by placing concrete from the hole bottom upward through water or drilling fluid using the *tremie process* described in Section 8.10. The concrete displaces some of the remaining solids which then float upward, suspended at the rising concrete surface. Although the shaft-to-rock load transfer is usually somewhat less than for hand-cleaned rock and dry-placed concrete shafts, high-capacity shafts have been placed by this procedure when the groundwater level has been 100 ft (30 m) above the rock surface. Obviously, such foundations cannot be inspected directly. Core borings in the shaft bottom before concreting can check the rock quality below the shaft bottom. Downhole geophysical logging through plastic casing set in the shaft before concreting has been used to verify the shaft concrete to rock contact and the quality of the concrete in the shaft. Seismic wave propagation from hammer blows or ultrasonic transducers on the hardened concrete shaft can detect large defects in the shaft concrete.

It is possible to drill through softer clays and even sandy overburden soils below the water table and into rock without casing (except for approximately 10 ft (3 m) at the top of the hole) by using drilling fluid. The fluid supports the hole walls and suspends the finer drill cuttings so they can be returned to the surface by pumping the drilling fluid. Using drilling fluid to support the hole walls requires skill and experience; there is a risk of cave-in, both during drilling and later during concreting. The greatest danger is from raising and lowering the drilling tools so rapidly that turbulence and pressure fluctuations are generated in the drilling fluid that cause soil in the hole wall to fall inward. The result is a loss of soil arching, which leads to more caving. It is impossible to inspect the shaft during concrete placement because the fluid fills the hole and only the top of the shaft can be observed. Therefore, the integrity, skill, and experience of the constructor are the keys to success.

An elaborate processing plant is necessary on the site to remove excess solids from the drilling fluid, to mix new fluid, and control its density and viscosity. This is costly and requires significant space on what is usually a crowded site. Eventually, the excess fluid must be disposed of in compliance with environmental regulations, which usually means a designated area at a significant distance from the site. This is inconvenient and expensive. However, many slurry shafts have been successful; a few have not. The difference is difficult to discern during construction. Borehole sensing through a plastic tube cast into the shaft and the ultrasonic sounding discussed in the previous section can detect large defects in the hardened concrete. Unfortunately, small defects that are not detected during construction may become obvious when the structure is fully loaded; at that time, remedies are difficult and expensive. Therefore, a conservative design and an experienced constructor are essential.

## 8.10 REINFORCING STEEL AND CONCRETE PLACEMENT

The open shaft becomes a foundation with the emplacement of the structural concrete in it. If the hole is dry, concrete can be dumped in the shaft from the ground surface, guided by a cylindrical chute. Experience has shown that there is no significant segregation of the concrete when it is dumped without a chute, provided the concrete does not bounce off the shaft sides and there is minimum reinforcing for it to impact on during falling, as reported by Reese and O'Neill (1988). Moreover, to some extent, the impact at the bottom helps to consolidate the concrete.

If there is much water in the shaft, dumped concrete mixes erratically with it, making it porous and weak. Therefore, it is necessary to place the concrete below water level by *tremie* (Fig. 8.15). The tremie is an 8- to 12-in. (200- to 300-mm) diameter pipe that extends from the ground surface to the

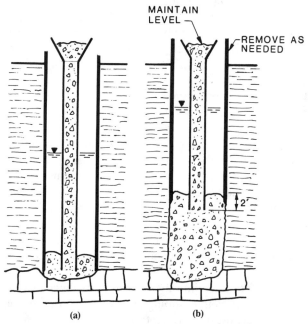

FIG. 8.15. *Tremie Concrete for a Water-filled Shaft: a. Beginning Concrete Flow into the Shaft Bottom; b. Raising the Casing and Tremie Tube as Concrete Fills the Shaft (Water Rises in the Casing and Should Be Removed When it is Above the Groundwater Level)*

hole bottom. At the top is a connection for directing the concrete into the tremie tube, either a funnel-shaped chute or a connection to a concrete pump. At the bottom of the tremie, there is a flap valve that keeps water out of the tremie tube when it is being lowered to the shaft bottom. Once the tremie rests on the bottom and the tremie tube is filled with concrete, it is raised above the bottom approximately 1 ft (300 mm), or slightly more depending on the shaft diameter. The flap valve opens, allowing the concrete to flow outward and fill the shaft and rise 2 to 3 ft (600 mm to 1 m) above the tremie bottom. Once the process has started, concrete is fed steadily into the tremie tube from above. The tremie is continuously raised during concreting. Otherwise the accumulating friction, plus some initial set, may cause the concrete flow to stop. However, because it is difficult to remove sections of the tremie as it is lifted, and because there may be air gaps introduced by stopping the pour to shorten the tremie tube, inexperienced crews sometimes gamble on removing the tremie later. However, it may not move, being held in the hardening concrete. Instead, the tremie should be steadily and carefully lifted so that its bottom remains immersed in the fresh concrete; otherwise the outward flowing concrete becomes contaminated by mixing with the water or drilling fluid that is being displaced. When this occurs, the integrity of the shaft is damaged by water or drilling fluid inclusions causing honeycombed or diluted, weak concrete.

The concrete mix is specially designed for tremie use. The coarse aggregate maximum size is 0.75 to 1 in. (19 to 25 mm) depending on the tremie diameter and the slump is 7 to 9 in. (180 to 230 mm). This means a higher cement content for a required strength and somewhat greater shrinking following curing. Therefore, load transfer by shear with large diameter shafts should be conservatively evaluated.

The tremie process requires both experience and diligence to produce good shaft concrete. If the pour is halted for a period sufficient for the concrete to begin to harden, the tremie is withdrawn and cleaned. The pouring is then resumed after roughening, and cleaning the concrete surface previously placed as a reasonable bond will be established between the two pours. If the shaft is not reinforced, a short reinforcing cage is sometimes inserted halfway into the concrete in the shaft immediately before the stoppage. This ensures structural continuity despite the cold joint caused by the delay.

Pumping the concrete into the shaft is sometimes used Instead, of the tremie. The same concrete pumping equipment is used for high-rise buildings. However, because the concrete is pumped downward, more care is necessary in order to keep the flow of concrete continuous to avoid air gaps in the delivery lines. The discharge line is raised during concrete placement, keeping the nozzle approximately 3 ft (1 m) below the concrete surface in the shaft.

When reinforcing steel is required for uplift, lateral load resistance of the shaft, or to augment shaft capacity, it is prefabricated and inserted in the shaft either before concrete placement begins or after placing approximately 1 to 3 ft (300 to 900 mm) of concrete in the shaft bottom. In this case, a concrete chute or a tremie is necessary. Any splices in the longitudinal steel are done preferably with mechanical couplings instead of overlapping the bars so as to minimize any interference between the steel and concrete movement during placement. The largest possible diameter bars are use to maximize the space between bars and make it possible for the concrete to flow easily through the reinforcing cage. Connectors are staggered, if possible, to minimize planes of interference with concrete flow.

When shaft uplift resistance is required, there are two alternatives. First, the load is transferred from the shaft reinforcing to the rock by shear in the same way that shear enhances the downward capacity. However, many limestone formations contain near-horizontal fissures, which have little or no tensile strength. In such cases, the upward resistance may be limited to that of an inverted cone of rock whose tip is at the base of the drilled shaft and whose radius is equal to the depth of the shaft below the rock surface. This is augmented by the weight of the cylinder of soil above the cone. If this is not sufficient, holes are drilled into the rock at the base of the shaft and filled with cement grout before shaft concreting is begun. The length of these holes is determined by the mass of rock to be engaged and the transfer of those forces to the anchor bars. Reinforcing bars sufficient to resist the uplift forces safely and transmit the required shear to the rock are set into the holes. The bars extend above the hole bottom sufficiently to transmit their forces to the reinforcing cage that is subsequently placed into the shaft.

Casing in the hole is raised slowly while the concrete is being placed, as the bottom of the casing is always more than 1 ft (300 mm) below the concrete surface. If it is lifted faster than the concrete rises, a cave-in of the exposed and unprotected shaft wall can impair the structural continuity of the shaft. If the casing is raised too slowly, the hardening concrete may bond to the casing. This can make it impossible to raise the casing further without raising the concrete in the casing with it. The setting concrete will pull apart, thus creating a void. A large, unfilled void in the shaft will cause settlement when the shaft is load. If the void should fill with soil that caved into the void from the now unsupported surrounding hole, the extrusion of the soil can possibly delay the settlement. Of course the resistance to any lateral load is decreased by a large concrete void. If the shaft is reinforced, the reinforcing in the gap can buckle under load. If the concrete hardens in the casing during a halt in concrete placement, it is better to abandon the casing and resume concrete placement despite the casing cost. Replacing a faulty foundation will be far more expensive.

In order to check the concrete for large voids, a plastic pipe can be placed in the shaft close to its center before concreting. It becomes embedded when the shaft is concreted. A downhole density logger (as discussed in Chapter 5) determines if large voids are present. Alternatively, sonic and ultrasonic impulses introduced at the top of the shaft and the reflected responses are measured. Interpretation of the response records can indicate if there are large defects in the shaft.

Because of the uncertainties, the design of shaft foundations supported on solutioned limestone is necessarily conservative. The construction sometimes is expensive. However, in many situations, it is the only way to build heavily loaded structures on a solutioned-limestone formation.

# CHAPTER 9

# RISK AND RISK ACCEPTANCE

## 9.1 DAMAGE AND CASUALTIES

In areas underlain by limestones, particularly in humid regions, damage to structures occurs from a sequence of events starting with limestone solution and followed by soil erosion. Two forms of ground movements predominate: 1) subsidence from continuing rock surface solution or from the early stages of erosion ravelling dome development and 2) sinkhole dropouts from soil dome collapse. Rock collapse is extremely rare compared to the life of man-made structures. Sometimes the damage is focused on a structure that provides an aggravating impetus for the event, such as a single-family dwelling with downspouts that saturate the ground adjacent to footings or an industrial plant with leaking pipes. Other subsidences and collapses may be the result of regional environmental disturbances such as groundwater lowering for quarry excavation, construction drainage, or for water supply.

The damage ranges from slow settlement of part of a house or building to collapse of part or all of a structure, as illustrated in some of the accounts of sinkhole problems in previous chapters. Occasionally there is personal injury; rarely there is loss of life. The lack of casualties is attributed to the warning of impending trouble given by ground surface fissures, the slow rate of movement of many failures, and, particularly, the relatively small area of each new subsidence or sinkhole. By way of contrast, property damage ranges from inconvenience to catastrophic loss.

When failures do occur in populated areas, they usually make headlines in local newspapers, radio, and TV news broadcasts. Because casualties are infrequent, the news is quickly forgotten unless there is some special human experience that catches the public attention. For example, three decades ago, a poor man, living alone in a simple cabin went to bed late at night after becoming partly intoxicated. He had attended a prayer meeting at his church where the preacher exhorted the congregation to resist the evils of alcohol and the place in Hell reserved for drunkards. His remorse for his chronic drinking led him to more drink after he returned home in order to forget his guilt. The next morning, he awoke to his alarm clock but found it was still dark. There was no light from the windows and no electricity. He stumbled to the front door, opened it only to find a wall of earth before him. In his befuddled condition, he thought that the devil had dragged him down into the earth along the passage to Hell. People on the street noted the hole in the ground, the reasonably intact cabin roof at about the ground level, and heard the man's screams about the devil dragging him deeper. Eventually, the fire department rescued him. The county demolished his broken house and filled the hole with earth. According to newspaper accounts, he never drank again. This little event received far more publicity than the several hundred other sinkholes in that county occurring during that same year and with most having far more property damage.

Like many such failures, there was no investigation of why it developed; there were no casualties and there was no property loss to his neighbors or to the public. The collapse could have been caused by natural geologic events, triggered by the lowering of the water table by a quarry nearby, or caused by the water dripping from the old man's roof onto the ground surface on both sides of his house.

## 9.2 THE RISKS ASSOCIATED WITH SINKHOLES

Risk is the expression of the overall chance of something bad happening. As such, it has at least three dimensions: 1) the nature of the occurrence including the damage, 2) how often it occurs, and 3) where it occurs. If these three aspects of the risk can be quantified, it should be possible to make rational decisions about how risk can be evaluated and eventually managed or reduced in the most cost-efficient manner. For subsidences and sinkholes, risk could be an expression of how many sinkholes occurred in a year in a particular area such as a city or county. One of the first such efforts to define the risk of sinkhole development was in Polk County, FL, an area of growing population at the same time when numerous new sinkholes developed. These occurred during a period when the groundwater table was depressed by water well pumping (see Fig. 1.1). A county officer responsible for emergency planning compiled information on the formation of new sinkholes. The locations and dates of their occurrences were displayed on a map in his office. The map was regularly examined by builders, real estate developers, and, eventually, agents who sold sinkhole insurance.

The Florida Sinkhole Institute was formed in 1982 at the University of South Florida, at Winter Park. Its formation was partially a response to the huge sinkhole that occurred in 1981 (Fig. 1.3). Among its activities, the

Institute compiled data on sinkhole occurrences in Florida in a detailed database. In areas where the data were reasonably complete over a sufficient period of time, studies were made to relate the frequency of their occurrence in specific areas to climate, geology, and groundwater conditions. This work has been described by Beck (1991). (Unfortunately, the Institute closed in 1992 because of lack of funding. The Florida Department of Natural Resources has taken over the database. Some of its other activities, such as sponsoring conferences on karst, have been continued by Dr. Beck, its former director, in his spare time.) Other organizations, such as state geological or natural resources departments and highway departments in the United States, have developed maps and databases on karst in specific areas where such features and problems occur. Similar efforts are underway in other parts of the world where solution subsidence and sinkholes occur. However, because of the variable quantity and quality of the data on the sinkhole locations, and particularly on why they developed, a quantitative evaluation of risk continues to be elusive.

A risk analysis needs information on four topics: 1) the failure and its consequences, 2) the location including geology and environment, 3) possible mechanisms, and 4) the date. This requires data obtained in a uniform manner over a sufficiently long period of time so that a reasonably good observed frequency of occurrence can be computed. The factors that must be included in the analysis are both partially natural and partially the result of human activity. Because human activity greatly influences sinkhole location and frequency, data valid for a particular area and time period may be misleading for future sinkhole activity. It is not likely that anything better than a subjective evaluation of the human factors can be made. Therefore, the sinkhole maps correlated with human activity can be of value in determining which areas are more prone to trouble than others. It is doubtful that a statistically sound model for any area can be developed without very great expense. Moreover, no estimates have been made that suggest that the money saved in the long run will exceed the cost of the data collection and evaluation that will be necessary for a reliable risk analysis.

It is likely that such economic arguments caused the closing of the Florida Sinkhole Institute despite its important work.

## 9.3 ACCEPTING RISK OF THE UNKNOWN

Although property owners and the public are unwilling to admit that they accept some risk, the building of cities, industries, and housing on areas underlain by limestone is silent testimony of the people involved that some risk is acceptable in exchange for constructing in such a region. Some segments of the public would like to believe that all risk can be eliminated thorough scientific studies, responsible engineering, and careful construction enforced by punitive laws. This myth is a form of wishful thinking. It is encouraged by those who profit from such claims, i.e., irresponsible politicians, sellers of "snake oil" cures, and those who profit from disaster.

Meanwhile, the public continues to occupy areas that are prone to sinkholes and related problems. They have apparently decided that the value of life in the area is greater than the cost associated with the risk. However, when failure does occur, many that are damaged want someone else to pay.

It is sometimes possible to force the idle dreamers to share in the risk if they wish to enjoy the benefits of living. Highway departments post signs in areas prone to rock falls that read, "Watch for falling rock." The message is that if one wishes to benefit from travel through that area that person must accept the risk. (What can you do to protect yourself if you see a rock falling toward you?)

A county in central Alabama, within a few miles of the Golly Hole (Fig. 3.14), experienced reoccurring sinkholes along a secondary road. When the cost of continuing to fill each new sinkhole depleted the county road budget to the extent that no other road maintenance was possible, the county engineer closed it to all vehicles. The public threatened to sue for the cost of a longer route between towns. The authorities answered by opening the road and posting warning signs at each end (Fig. 9.1). The travelling public traded the convenience of using the road for the risk of a sinkhole swallowing their vehicle. The risk is very small, because the road is not heavily travelled, but it is real. Apparently, there have been no serious casualties. As far as the author knows, this is the only such warning devoted to sinkholes.

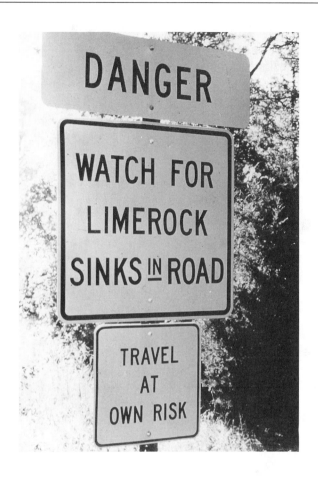

FIG. 9.1.   *Highway Sign Warning of Sinkhole Risk, Shelby County, AL*

# BIBLIOGRAPHY AND REFERENCES

Beck, B. F. (1991). "On Calculating the Risk of Sinkhole Collapse." Appalachian Karst, *Proc., Appalachian Karst Symposium,* E. H. Kastning and K. M. Kastning, eds., National Speleological Society, Inc., Huntsville, AL.

Chan, S. F. (ed.) (1986). *Foundation Problems in Limestone Areas of Peninsular Malaysia,* Geotech. Eng. Div., The Institution of Engineers, Malaysia, Kuala Lumpur.

Chen, F. H. (1975). *Foundations and Expansive Soils.* Elsevier Scientific Publishing Co., Amsterdam, The Netherlands.

Cooper, A. H. (1995). "Subsidence hazards due to the dissolution of Permian gypsum in England." *Karst Geohazards,* Beck. B. F., ed., A. A. Balkema, Rotterdam, The Netherlands, 23–31.

DeStephen, R. A., and Wargo, R. H. (1992). "Foundation Design in Karst Terrain." *Bulletin of the Association of Engineering Geologists,* 39(2), 165–173.

Fairbridge, R. W. (1968). *Encyclopedia of Geomorphology.* Reinhold Book Co., New York, NY.

Ford, D. C., and Williams, P. W. (1989). *Karst Geomorphology and Hydrology.* Unwin Human, Ltd., London, England.

Getchell, F. J. (1995). "Subsidence and related features in the Tully Valley, central New York." *Karst Geohazards,* Beck, B. F., ed., A. A. Balkema, Rotterdam, The Netherlands, 31–42.

Goodman, R. E. (1988). *Introduction to Rock Mechanics.* John Wiley & Sons, New York, NY.

Gray, M. (1974). McAfee, R., Jr., Wolf, C. L., ed., *Glossary of Geology.* American Geological Institute, Washington, D.C.

Heath, W. E. (1995). "Drilled pile foundations," *Karst Geohazards,* Beck, B. F., ed., A. A. Balkema, Rotterdam, The Netherlands, 371–375.

House, J. K., (1995). "Carbonate rock investigation guidance policy for siting landfills in Tennessee," *Karst Geohazards,* Beck, B. F., ed., A. A. Balkema, Rotterdam, The Netherlands, 511–516.

Hill, C. A. (1987). *Geology of Carlsbad Cavern and Other Caves in the Guadalupe Mountains, New Mexico,* Bulletin 117, New Mexico Bureau of Mines and Mineral Resources, Albuquerque, NM.

Hyatt, J. A., and Jacobs, P. M. (1995). "Recent Sinkhole Development on the Dougherty Plain at Albany, Ga.," *Paleogene Carbonate Facies and Paleeogeography of the Dougherty Plain Region.* Georgia Geological Society Guidebooks, Ga., 15(1).

Jamal and Associates (1982). The Winter Park Sinkhole. Jamal Associates, Inc., Winter Park, FL. (A report to the City Commission, city of Winter Park, FL.)

Jennings, J. E., Brink, A. B. A., Louw, A., and Gowan, G. D. (1965). "Sinkhole and subsidences in the Transvaal dolomite of South Africa." *Proceedings of* the 6th *International Conference on Soil Mechanics and Foundation Engineering*, University of Toronto Press, Toronto, Canada, 51–54.

Jennings, J. E. (1971). *Karst*, MIT Press, Cambridge, Mass.

Jennings, J. N. (1985). *Karst Geomorphology*. Basil Blackwell, Oxford, England.

Jennings, J. N. (1983). "Karst landforms," *American Scientist*, 71, 578–588.

Julian, H. E., and Young, S. C. (1995). "Conceptual model of groundwater flow in a mantled karst aquifer and the effects of the epikarst zone," *Karst Geohazards*, Beck, B. F., A. A. Balkema, Rotterdam, The Netherlands, 131–141.

LaMoreaux, P. E., Raymond, D., and Joiner, T. J. (1978). *Hydrology of Limestone Terranes—An Annotated Bibliography of Carbonate Rocks*. Geological Survey of Alabama, University, AL.

Milanovic P. (1981). *Karst Hydrology*. Water Resources Publications, Littleton, CO.

Kauschinger, J. L., and Welsh, J. P. (1989). "Jet Grouting for Urban Construction," *Design, Construction and Performance of Deep Excavations in Urban Construction, Proceedings of 1989 Seminar*, Boston Society of Civil Engineers, ASCE, New York, NY.

Knott, D. L., Rojas-Gonzales, L., Newman, F. B., (1993). *Guide for Foundation Engineering in Pennsylvania Karst*, Commonwealth of Pennsylvania Department of Transportation, Research Project 90-12, GAI Consultants, Inc., Monroeville, PA.

Quinlan, J. F., and Ewers, R. O. (1985). "Groundwater flow in limestone terranes: strategy rationale and procedures for reliable, efficient monitoring of groundwater quality in karst areas." *Proceedings, 5th National Symposium and Exposition on Aquifer Restoration and Groundwater Monitoring*, National Water Well Association, Worthington, OH, 197–234.

Reese, L. C., and O'Neill, M. W. (1988). *Drilled shafts: Construction, Procedures and Design Methods*, U.S. Department of Transportation, Federal Highway Administration Office of Implementation, McLean, VA and ASDC: The International Association of Foundation Drilling, Dallas, TX.

Sams, Clay, and Moshen Sefat (1995). "Analysis of Surface Subsidence by Pinnacle Punching." *Report on Residual Soil Settlement*, Law Engineering and Environmental Services, Charlotte, NC.

Siegel, T. C., and Belgerie, J. J. (1995). "The importance of a model in foundation design over deeply weathered pinnacled carbonate rock," *Karst Geohazards*, Beck, B. F., ed., A. A. Balkema, Rotterdam, The Netherlands, 375–382.

Sitar, N. (ed.) (1988). *Geotechnical Aspects of Karst Terrains*, Geotechnical Special Technical Publication No. 14, ASCE, New York, NY.

Smith, D. L., and Atkinson, T. C. (1976). "Process, landforms and climate in limestone regions." *Geomorphology and Climate*. John Wiley & Sons, New York, NY, 367–409.

Sowers, G. F. (1979). *Introductory Soil Mechanics and Foundation Engineering*. Macmillan Publishing Co., New York, NY.

Sowers, G. F. (1975). "Failures in limestones in humid subtropics." *Journal of the Geotechnical Engineering Division*, ASCE, 101(GT8), 771–787.

Sowers, G. F. (1976). "Foundation bearing in weathered rock." *Proceedings, Specialty Conference on Rock Engineering for Foundations and Slopes,* ASCE, New York, NY, Vol. 2.

Sowers, G. F. (1975). "Settlement in terrains of well-indurated limestone." *Proceedings, Conference on the Analysis and Design of Building Foundations,* Lehigh University, Bethlehem, PA, 701.

Sowers, G. F. (1976). "Mechanisms of Subsidence due to Underground Openings." *Subsidence Over Mines and Caverns, Moisture and Frost Actions and Classification; Transportation Research Record 612,* Transportation Research Board, National Academy of Sciences, Washington, DC.

Sowers, G. F. (1981). "Rock permeability or hydraulic conductivity—an overview." *ASTM Special Technical Publication 746,* ASTM, Philadelphia, PA.

Sowers, G. F. (1984). "Correction and Protection in Limestone Terrain." *Proceedings of the First Multi-disciplinary Conference on Sinkhole,* Florida Sinkhole Institute, University of Central Florida, Orlando, FL. Also published by *Environmental Geologic Water Science* (1988). 8(1/2), 77–82.

Sweeting, M. M. (1972). *Karst Landforms,* Macmillan, London, England.

Terzaghi, K., (1948). "Rock defects and loads on tunnel supports." *Rock Tunnelling with Steel Supports.* Commercial Shearing and Stamping Co., Youngstown, OH. Also in Terzaghi, K., and Peck, R. B. (1949). *Soil Mechanics in Engineering Practice,* John Wiley & Sons, New York, NY.

White, W. B. (1988). *Geomorphology and Hydrology of Karst Terrains.* Oxford University Press, Oxford, England.

White, W. B., and White, E. L. (1995). "Thresholds for soil transport and the long term instability of sinkholes." *Karst Geohazards,* Beck, B. F., ed., A. A. Balkema, Rotterdam, The Netherlands, 73–79.

Woods, R. D. (ed.) (1994). *Geophysical Characterization of Sites.* ISSMFE Technical Committee 10, Oxford and IBHI Publishing Co. Pvt. Ltd, New Delhi, India.

# INDEX